Experiments in General Chemistry: Inquiry and Skill Building

THIRD EDITION

Vickie Williamson
Texas A&M University

Larry Peck
Texas A&M University

Kathleen McCann
University of Pikeville

Australia • Brazil • Mexico • Singapore • United Kingdom • United States

*Experiments in General Chemistry:
Inquiry and Skill Building,* **Third Edition**
**Vickie Williamson, Kathleen McCann and
Larry Peck**

Product Director: Dawn Giovanniello

Product Manager: Lisa Lockwood

Content Developer: Peter McGahey

Product Assistant: Nellie Mitchell

Marketing Manager: Janet Del Mundo

Manufacturing Planner: Becky Cross

Design Director: Jack Pendleton

Cover Image: © Shutterstock.com/creativenv;
© iStock.com/kali9

Production Management, and Composition:
Lumina Datamatics, Inc.

Intellectual Property:

   Analyst: Christine Myaskovsky

   Project Manager: Kathryn Kucharek

For product information and technology assistance, contact us at
Cengage Customer & Sales Support, 1-800-354-9706

For permission to use material from this text or product,
submit all requests online at **www.cengage.com/permissions.**
Further permissions questions can be e-mailed to
permissionrequest@cengage.com

Library of Congress Control Number: 2017959786

Student Edition:
ISBN: 978-1-337-39924-1

Cengage
20 Channel Center Street
Boston, MA 02210
USA

Cengage is a leading provider of customized learning solutions with employees residing in nearly 40 different countries and sales in more than 125 countries around the world. Find your local representative at
www.cengage.com.

Cengage products are represented in Canada by Nelson Education, Ltd.

To learn more about Cengage platforms and services, visit **www.cengage.com.** To register or access your online learning solution or purchase materials for your course, visit **www.cengagebrain.com.**

Printed in the United States of America
Print Number: 01   Print Year: 2018

Contents

To: Instructors and Students

A researcher may design a chemistry experiment with the goal of gathering data that supports or refutes the researcher's hypothesis. In another experiment, perhaps in the same laboratory, a researcher may carefully follow a procedure described in the literature in hopes of developing knowledge and skills that will influence future experiments. In this manual, we have divided the experiments into three categories: Skill Building experiments, Guided Inquiry experiments, and Open Inquiry experiments.

Skill Building experiments are highly structured and direct students through activities. One sometimes finds this style of experiment referred to as "Cookbook Experiments"; however, Skill Building experiments are often the most efficient way to teach skills that may be needed in future experiments or to demonstrate previously developed concepts. While many instructors are accustomed to this type of experiment and while it can be an efficient use of time, it is not typical of all research laboratory activities.

In Guided Inquiry experiments, students collect data/observations on designated variables without previously studying the relationship between variables. Students are guided to the logical organization, comparison, analysis, and interpretation of data. Students use their data to come to a generalization about the variables. The processes of forming generalizations or hypothesizing so the explanation "fits" the data and testing the hypothesis are important aspects of this type of laboratory.

Open Inquiry experiments ask students to design experiments to answer questions based on their understanding of the concept(s). Students choose the variables, procedure, and the exact nature of the question/hypothesis.

Each type of experiment has advantages and disadvantages. When the skill or end product is the primary goal, a Skill Building experiment is used. When it is possible for the students to design their experiment to test a hypothesis or concept, an Open Inquiry experiment is used. When the students need some direction on a topic or concept, the Guided Inquiry experiment is used. This manual contains a mix of the three types.

We believe that with a combination of experiment styles, the information and skills gained by the students can be maximized while the amount of time spent in the lab can be minimized. We put a heavy emphasis on the particulate nature of matter, often asking students to draw what is happening to the particles during a reaction or process, believing that this will aide the student's conceptual understanding of the chemical phenomena. We trust that you will find these experiments well designed, informative, efficient, instructional, and enjoyable.

Vickie M. Williamson, M. Larry Peck, and Kathleen McCann

To: Instructors:

All labs in this third edition are updated and revised with the input from the multiple adoptions at both large and small institutions. A new lab on intermolecular forces is included, along with safety agreements for two semesters. The instructor's manual has been updated and revised. It includes a recommended Pre-Lab discussion, sample answers to the Pre-Lab exercises and Post-Lab questions, and a list of needed chemicals and equipment for each experiment.

Acknowledgments

We would like to acknowledge the students at the University of Pikeville for testing our new additions to this edition. We also need to thank Dr. Elmo J. Mawk, Texas A&M University, for his help in editing with the original manual.

Laboratory Safety

A chemical laboratory can be a hazardous place to work if common safety rules are not enforced. If basic rules are strictly enforced, the chance of one being injured becomes very small. In this course, experiments that involve some of the safer chemicals and equipment have been selected. However, you will still need to use electric equipment, hot water, and concentrated solutions. These are only safe to work with if you follow the correct procedures. With proper understanding of what you are doing, careful attention to safety precautions, and adequate supervision, you will find the chemical laboratory to be a safe place in which you can learn much about chemistry.

Laboratory accidents belong to two general categories of undesirable events: mishaps caused by your own negligence and accidents beyond your control. Although accidents in the laboratory fortunately are rather rare events, you nevertheless must be familiar with all safety rules and emergency procedures. If you know and follow safe working practices, you will pose no threat of serious harm to yourself or others.

Every laboratory system will have some unique features for the prevention of accidents or for handling emergencies. It is important that you become thoroughly familiar with the specific safety aspects of your own laboratory area. Some general precautions and procedures applicable to any chemical laboratory are summarized below.

SAFETY

1. **WEAR APPROVED EYE PROTECTION AT ALL TIMES.** Very minor laboratory accidents, such as the splattering of solution can cause permanent eye damage. Wearing laboratory goggles can prevent this eye damage. In the chemistry teaching laboratories, safety glasses (goggles) of an **approved** type **must be worn** by all persons in the room at all times that anyone is working with or transporting glassware or conducting any experimental work. Experimental work includes simple tasks such as transporting chemicals or glassware, obtaining quantitative measurements that involve non-sealed containers, etc. Light-weight "visitors' shields" or prescription glasses with side shields are acceptable only for laboratory visitors if your institution permits them, but are not suitable for routine laboratory work.

2. **WEAR PROPER PROTECTIVE CLOTHING.** Proper protective clothing **must** be worn by all persons in the room at all times that anyone is working with or transporting glassware or conducting any experimental

work. Exposed skin is particularly susceptible to injury from splattering of hot, caustic, or flammable materials. Students and instructors need to be protected from their necks to below their knees. This requirement includes **no shorts, no short skirts, no sleeveless garments,** and **no bare midriffs.** Long lab coats or aprons are required if shorts or short skirts are worn. Makeshift coverage such as shirts being used as aprons, paper taped over the knees, etc., is not considered to be suitable. Tight fitting clothing, long unrestrained hair, clothing that contains excessive fringe or even overly loose-fitting clothing may be ruled to be unsafe.

3. **WEAR PROPER PROTECTIVE FOOTWEAR.** No sandals, no open-toed shoes, and no foot covering with absorbent soles are allowed. Any foot protection that exposes any part of one's toes is unsuitable for wear in the laboratory.

4. **NEVER EAT, DRINK, SMOKE, OR VAPE IN A CHEMICAL LABORATORY.** Tiny amounts of some chemicals may cause toxic reactions. Many solvents are easily ignited. Food and drinks are never allowed in the labs. This includes all visible insulated water bottles or mugs, containers of water or flavored drinks, containers of ice intended for consumption, etc. If a food or drink container is empty or unopened, it needs to be inside a backpack, etc., and out of sight.

5. **NEVER WORK IN A CHEMICAL LABORATORY WITHOUT PROPER SUPERVISION.** Your best protection against accidents is the presence of a trained, conscientious supervisor, who is watching for potentially dangerous situations and who is capable of properly handling an emergency.

6. **NEVER PERFORM AN UNAUTHORIZED EXPERIMENT.** "Simple" chemicals may produce undesired results when mixed. Any experimentation not requested by the laboratory manual or approved by your instructor may be considered to be unauthorized experimentation.

7. **NEVER INHALE GASES OR VAPORS UNLESS DIRECTED TO DO SO.** If you must sample the odor of a gas or vapor, use your hand to waft a small sample toward your nose.

8. **EXERCISE PROPER CARE IN HEATING OR MIXING CHEMICALS.** Be sure of the safety aspects of every situation in advance. For example, never heat a liquid in a test tube that is pointed toward you or another student. Never pour water into a concentrated acid. Proper dilution technique requires that the concentrated reagent be slowly poured into water while you stir to avoid localized overheating.

9. **BE CAREFUL WITH GLASS EQUIPMENT.** Cut, break, or fire-polish glass only by approved procedures. If a glass-inserter tool is not available, use the following procedure to insert a glass rod or tube through a rubber or cork stopper. Lubricate the glass and the stopper, protect your hands with a portion of a lab coat or a towel, and use a gentle twisting motion to insert the glass tube or rod.

10. **NO REMOVAL OF CHEMICALS OR EQUIPMENT FROM THE LABORATORY.** The removal of chemicals and/or equipment from the laboratory is strictly prohibited and is grounds for severe disciplinary action.

11. **NO HORSEPLAY.** Horseplay and pranks do not have a place in instructional chemistry laboratories.

12. **NO BICYCLES, ROLLER-BLADES, ETC.** Bicycles are not allowed in the buildings where chemistry labs meet. Using hoverboards, skate boards, in-line skates, roller-skates, and unicycles is also not allowed. If skates, etc., are brought inside the building, they must be stored where they will not be in anyone's work area or in any traffic area.

13. **NEVER PIPET BY MOUTH.** Always use a mechanical suction device for filling pipets. Reagents may be more caustic or toxic than you expect.

EMERGENCY PROCEDURES

1. **KNOW THE LOCATION AND USE OF EMERGENCY EQUIPMENT.** Find out where the safety showers, eyewash spray, and fire extinguishers are located. If you are not familiar with the use of emergency equipment, ask your instructor for a lesson.

2. **DON'T UNDERREACT.** Any contact of a chemical with any part of your body may be hazardous. Particularly vulnerable are your eyes and the skin around them. In case of contact with a chemical reagent, wash the affected area immediately and thoroughly with water and notify your instructor. In case of a splatter of chemical over a large area of your body, don't hesitate to use the safety shower. Don't hesitate to call for help in an emergency.

3. **DON'T OVERREACT.** In the event of a fire, don't panic. Small, contained fires are usually best smothered with a pad or damp towel. If you are involved in a fire or serious accident, don't panic. Remove yourself from the danger zone. Alert others of the danger. Ask for help immediately and keep calm. Quick and thorough dousing under the safety shower often can minimize the damage. Be prepared to help, calmly and efficiently, someone else involved in an accident, but don't get in the way of your instructor when he or she is answering an emergency call.

These precautions and procedures are not all you should know and practice in the area of laboratory safety. The best insurance against accidents in the laboratory is thorough familiarity and understanding of what you're doing. Read experimental procedures before coming to the laboratory, take special note of potential hazards, and pay particular attention to advice about safety.

Take the time to find out all the safety regulations for your particular course and follow them meticulously. Remember that unsafe laboratory practices endanger you and your neighbors.

If you have any questions regarding safety or emergency procedures, discuss them with your instructor. Then sign and hand in the following safety agreement.

Safety Agreement

I have studied, I understand, and I agree to follow the safety regulations required for this course. I have located all emergency equipment and now know how to use it. I understand that I may be dismissed from the laboratory for failure to comply with stated safety regulations.

Signature _____

Print Name _____

Date   _____

Course   _____

Section   _____

Instructor _____

Person(s) who should be notified in the event of an accident:

_____

_____

_____

_____

_____

Signature _____

Local Address   **Permanent Address**

_____   _____

_____   _____

_____   _____

Phone Number _____   Phone Number _____

Email Address _____

Safety Agreement

I have studied, I understand, and I agree to follow the safety regulations required for this course. I have located all emergency equipment and now know how to use it. I understand that I may be dismissed from the laboratory for failure to comply with stated safety regulations.

Signature _____

Print Name _____

Date _____

Course _____

Section _____

Instructor _____

Person(s) who should be notified in the event of an accident:

_____

_____

_____

_____

_____

Signature _____

Local Address

_____

_____

_____

Phone Number _____

Email Address _____

Permanent Address

_____

_____

_____

Phone Number _____

Are Labels Accurate or Precise?

An Open Inquiry Experiment

INTRODUCTION

In non-quantitative terms we say that a value is accurate if it differs from a true value by less than some acceptable amount. We say that a value is precise if it is reproducible to some predetermined number of significant figures. Is the amount indicated on a package of a commercial product accurate and/or precise? Data Treatment and Graphing Data are discussed in the Common Procedures and Concepts Section at the end of this manual.

OBJECTIVES

During this experiment, you will design experimental procedures that can be used to investigate a question, to gain experience in working with the analytical balance, to gain knowledge in data treatment and the reporting of data, and to reinforce your understanding of the terms "accuracy" and "precision."

CONCEPTS

This experiment uses the following concepts: mass, average, standard deviation, percent deviation, intensive and extensive properties, systematic and random errors, accuracy, and precision.

TECHNIQUES

Using the analytical balance and transferring solids are just some of the techniques encountered in this experiment.

ACTIVITIES

You will have five small packets of sweetener (or similar product). Only one of the packets can eventually be opened. From an understanding of data treatment and random versus systematic errors, you are to design procedures that will allow you to calculate the average mass and the standard and percent deviation of the mass of the contents in the unopened packets. In your experimental design, you are to specify number and type of weighings needed in your experiment. Using your determination of the mass of the contents of one packet, you are to assume that the mass of the contents will have the same percent deviation as the mass of the unopened packets, and you are to also assume that the mass of the paper packet (packaging material) is the same for all five packets. From your knowledge of the values determined for the contents of the packets, you are to comment on the accuracy and precision of the mass indicated on the label of the packets.

CAUTION	

Do not taste any material found in the chemistry laboratory. Dispose of materials as directed by your instructor. Wear approved eye protection.

PROCEDURES

Experimental Design

1-1. In many experiments the data to be collected and the chemicals to be used are specified. In this experiment, you are to design the exact experimental procedures. This type of lab is often referred to as an open inquiry lab. "Open" means that you design the procedures.

1-2. Design an experiment to determine the average mass and the standard and percent deviation of the contents of five packets of sweetener. Remember that you can open only one packet, but you can assume that the mass of the contents of each packet will have the same percent deviation as the mass of the opened packet. You can also assume that the mass of the paper packet (packaging material) is the same for all five packets.

The design must include:

Problem Statement: This includes a few sentences describing the specific question(s) you are trying to answer with your experiment.

Proposed Procedures: This section contains the materials and equipment that you will use, the type of data you will collect (the variables you will measure), and the number of trials you are proposing. You should discuss safety considerations. Your planned experimentation should take up ⅔ of the lab period.

1-3. Your instructor will supply feedback on your experimental design BEFORE you do the experiment.

Experimentation

2-1. Make any changes that your instructor suggests, and then proceed to collect data. Record the data in your notebook and get your instructor's signature.

Report

3-1. After completing the experiment, you will write a lab report. Although you collect data and share ideas with a partner, you will be expected to write the final lab report independently. Your grade will depend on the thoroughness of your investigation, the presentation of your data, the careful analysis of the data, and the logic used to give reasonable results and explanations.

3-2. The lab report MUST include the following sections:

Problem Statement: This includes a few sentences describing what specific questions you are trying to answer with your experiment.

Procedures: This section contains the materials and equipment that you actually used, the type of data collected (the variables measured), and the number of trials done. Remember to discuss safety considerations. Your experimentation should take up ⅔ of the lab period.

Data/Analysis: Include the data you collected. When possible, data should be in tables with easy-to-read labels. Analysis of the data should also be included. (Analysis is what your data tells you.) Graphs (with labels, units, and titles), mathematical relationships, chemical equations, and algebraic equations should be given, and the connection to the data should be shown.

Conclusion: This is the generalization or explanation you have deduced from your experiment. You are to comment on the accuracy and precision of the mass indicated on the label of the packets. (For this you will need to have recorded the mass of the contents as indicated on the label or the amount indicated by your instructor.) This is also the place to make explanations for any data results that are counter to logical chemical ideas and to describe how you would change the experiment if you repeated it.

EXPERIMENT 1: ARE LABELS ACCURATE OR PRECISE?

Prelab Exercises

1. In your own words, define the following:

   a. mass

   b. average

   c. extensive property

2. Discuss the differences between accuracy and precision.

3. Give the problem statement for this experiment.

4. Propose procedures for this experiment (insert more sheets if necessary).

Date _____ **Student's Signature** _____

Instructor's Approval and Comments (to be added later)

Date _____ **Instructor's Signature** _____

1   EXPERIMENT 1:   ARE LABELS ACCURATE OR PRECISE?

Report Form

DATA

Data are collected in your notebook.

Date _____   **Instructor's Signature** _____

LAB REPORT

Attach this sheet to your lab report that includes the PROBLEM STATEMENT (actual), PROCEDURES (actual), DATA/ANALYSIS, and CONCLUSION.

POSTLAB QUESTIONS

1. Is the "label" mass true? On the average, did the manufacturers fill the packets just right? Did they overfill? Did they add less than claimed on the label? If a ±5% variation is acceptable, is the "label" mass true?

2. How would you modify this experiment if you were examining the labeled total contents of cans of a soft drink and could only open one can?

3. How would a chemist decrease the value of his/her standard deviation for a series of measurements performed in the laboratory?

4. Describe and give an example of a systematic error.

5. Answer any questions assigned by your instructor.

Date _____ **Student's Signature** _____

Density Measurements

A Skill Building Experiment

INTRODUCTION

Some of your earliest scientific observations may have been the floating of ice on water, the floating of oil on water, or the rising of a hot air or helium balloon. As you developed your scientific expertise, you learned to explain these and other observations using the concepts of density and density differences. Density is a very useful physical property that can be associated with a sample's buoyancy, temperature, purity, or identity. You may be able to think of other simple phenomena that involve density.

Density of solids and liquids is usually expressed as mass in grams per 1.00 mL of the liquid or solid. Density of gasses is usually expressed in grams per 1.00 L of a gas. Since it does not depend on the amount of matter present, density is an intensive property. A formula for the density of solids and liquids is:

$$\rho = \frac{m}{V}$$

where ρ = density (g mL^{-1}), m = mass (g), and V = volume (mL).

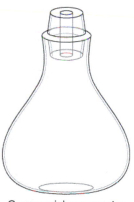

Commercial pycnometer

A classical method to determine the density of a liquid is to use a pycnometer (pik-nom-i-ter). One such pycnometer is shown. It is a small glass vessel with a glass stopper with a hole in it. If the vessel is filled and the stopper inserted, a small amount of liquid will be forced out the hole in the glass stopper. The excess liquid is removed with a towel; the volume of liquid in the vessel will be the same each time it is filled. The pycnometer is weighed empty and full. The mass difference equals the mass of the liquid. By knowing accurately the volume of the pycnometer and the mass of its contents, the density of the contents can be accurately and precisely calculated.

OBJECTIVES

During this experiment, you will gain experience in working with hot glass and the analytical balance, gain knowledge in data treatment and the reporting of data, and reinforce your understanding of some basic and frequently encountered laboratory concepts and techniques.

CONCEPTS

This experiment uses the following concepts: mass, volume, density, average, standard deviation, percent deviation, intensive and extensive properties, chemical and physical properties, systematic and random errors, and accuracy and precision. (See the Common Procedures and Concepts Section at the end of this manual for additional information.)

TECHNIQUES

Lighting and adjusting a burner, cutting glass tubing, working with hot glass, using the aspirator to obtain reduced pressure, and disposal of sharp items are just some of the techniques encountered in this experiment.

ACTIVITIES

You will make a pycnometer from a glass Pasteur pipet, use water of known density to determine the volume of your pycnometer, use your pycnometer to determine the density of an "unknown" liquid, and calculate the averages and the standard and percent deviations for your results.

CAUTION	

You will be handling sharp and hot glass that can cause severe cuts or burns if not properly handled. Hot glass looks just like cold glass. Common sense will help one to avoid injuries to oneself or to other students or the damage of property. Do not lay glass on the bench top or on any combustible surface until you are sure that it is cool. Place only cooled waste glass into the container designated for "glass disposal" or "sharps disposal." Fire-polish sharp edges on glass pieces when possible. (See Common Procedures and Concepts Section at the end of this manual.)

PROCEDURES

Construction of a Pycnometer

1-1. Light and adjust a Bunsen burner. Experiment with the adjustment of the burner's air intake and needle valve until you obtain the maximum opening of the needle valve and air vent that still produces a flame with the distinct bright-blue inner cone. (See Common Procedures and Concepts Section at the end of this manual.)

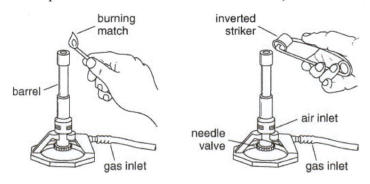

CAUTION

You will be working with hot flames and glass. Hot objects look no different than they look when cold. Avoid burns by hesitating before you touch anything that might be hot.

1-2. Your instructor will give you a glass Pasteur pipet. Hold it near the wider, non-constricted end with one hand. Place the other hand on the tapered portion of the pipet (near the center of the pipet). Holding the pipet with both hands, practice rotating the tube at a constant rate with both hands. Heat a spot on the non-constricted portion of the pipet that is approximately 2 cm (less than 1 inch) from where the taper begins. Rotate the pipet in the flame while keeping the blue inner cone portion of the flame just below the area being heated. Continue to heat and rotate the pipet until the glass begins to sag. DO NOT STOP!

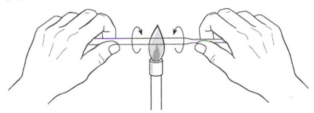

Heating must be continued to assure that the glass is sufficiently heated. Continue the heating and steady rotation. You should continue the heating and rotation until you find you are able to push the two cold ends with slight pressure toward the soft portion. This will prevent thinning of the soft glass. If the glass tube is not rotated and heated properly, the soft part might become twisted. DON'T LET IT TWIST!

1-3. When the selected portion of the pipet finally becomes very soft, remove it from the flame and pull outward on each end. The pulling needs to draw the soft part of the tube into a thin section (capillary) that is 6 to 8 cm length and approximately 1 mm inside diameter. Hold the tubing in a vertical position. Hold the top end of the glass tubing while allowing the other end to swing in the air. Allow the glass to cool slowly in the air. Slow cooling will avoid stress that may cause the glass to break later. While the glass is cooling, turn off your burner.

> ### WARNING
>
> It may take several minutes for the pipet to cool and to become safe to touch. The glass will cool more quickly where it is thinner. Not all areas will be equally hot. It may take a section of the glass tube several minutes to cool. Do not put the hot glass on the bench. Hold it until all sections have cooled.
>
> If you pulled the hot pipet too fast, too strongly, or too far, the capillary produced will have a diameter that is too small. Don't become discouraged if your first attempt is a less than perfect product. Try again. Practice is needed when working with glass.

1-4. Check carefully that your glass tube is cool. Use a file to make one scratch near the middle of the new capillary section of the pipet.

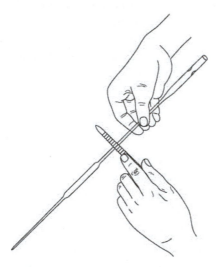

Support the tube on the left and right side opposite the scratch with your thumbs. You may wish to protect your fingers by placing a towel between your fingers and the glass. Break the capillary at the scratch by bending away from the scratch and pulling apart.

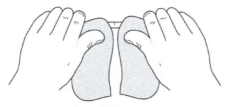

You should now have two pieces of glass: the body of the pycnometer plus a capillary section on one end and the constricted portion of the pipet on the other end. The constricted portion of the original pipet may be discarded or used in the following manner.

> ### WARNING
>
> Place only cooled waste glass into the container designated for "glass disposal" or "sharps disposal."

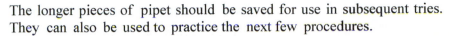

The longer pieces of pipet should be saved for use in subsequent tries. They can also be used to practice the next few procedures.

1-5. Re-light the Bunsen burner. Slowly close the air intake and needle valve until you obtain the smallest possible cool yellow flame. Practice bending the capillary by using the piece of waste glass produced in Procedure 1-4. To bend the capillary, carefully turn it in the cool flame until it is soft enough to bend into a "V" with a 30° to 40° angle. If the capillary tube becomes too soft, it could close. If this happens when you are making your pycnometer, you will have to start over at Procedure 1-1.

1-6. After you have successfully made the practice bend, use the same technique to bend a "30° to 40° V" in the long capillary but near the pycnometer body. Make a second bend in the capillary approximately 2 cm from the first bend. The last part of the capillary (~2 cm) should be parallel to the long axis of the pycnometer body. The pycnometer should resemble a Z. The correct shape of the finished pycnometer is shown.

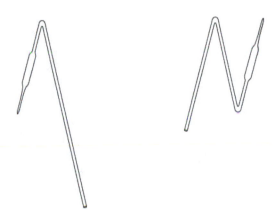

1-7. Using the file again, make one scratch mark on the originally constricted side of the pipet about 1.5 cm from the pycnometer's body. A single scratch is sufficient. Multiple scratches may yield a jagged break. Place your thumbs on the side opposite the scratch. Break the tubing by bending away from the scratch and pulling apart.

1-8. Fire-polish the two ends of the pycnometer by holding them very briefly in the cool flame but do not allow the openings to close. Heat the two ends until the openings are about as small as the diameter of the wire in a paper clip (~1 mm).

It usually takes more than one attempt to construct a usable pycnometer.

**Determination of the
Volume of the Pycnometer**

2-1. Take a watch glass and your pycnometer to the analytical balance. Place the watch glass on the balance pan and tare the balance and watch glass. Place the pycnometer on the tared watch glass. Record the mass of the pycnometer to the nearest 0.0001 g. Remove the pycnometer, re-tare the balance and watch glass, and reweigh the pycnometer. Repeat for a total of three weighings.

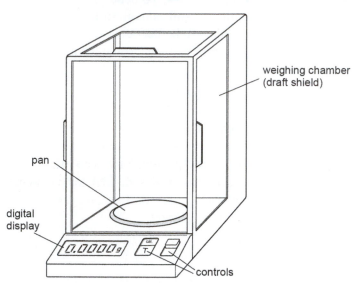

weighing chamber
(draft shield)

pan

digital
display

0.0000 g

controls

2-2. Go back to your work area; fill a small test tube with distilled water. Obtain a thermometer from your instructor. Record the temperature of the water (see comments after Procedure 2-5). The temperature of the water should be the same as the temperature of the room. If not, let the test tube and water sit about 5 minutes. The temperature of the room may be provided.

2-3. Place the top of the "Z" of the Z-shaped pycnometer into the test tube of water and tip the assembly until the pycnometer body is below the liquid level in the test tube. The pycnometer should spontaneously but slowly fill with water. Inspect the ends of the pycnometer to ensure that it is filled with liquid. Carefully wipe off any liquid clinging to the outside. The filled pycnometer must always be in a flat, horizontal position. If you tilt the pycnometer, some liquid may run out. Lay the filled pycnometer on the watch glass. Return to the analytical balance used in Procedure 2-1 and re-tare the balance with the watch glass on the balance pan. Weigh the filled pycnometer to 0.1 mg (0.0001 g) on the tared watch glass.

2-4. To empty the pycnometer, use a vacuum flask equipped with a rubber stopper that has one or two small holes. Carefully place the straight end of the pycnometer into the hole and turn on the aspirator. Dry the pycnometer by letting air pass through the pycnometer for a few minutes. (If the rubber stopper has a second hole in it, cover the second hole with your finger while the pycnometer is being emptied or dried.)

Table 2.1 *Density of water,*
ρ^T, g mL^{-1}

°C	g/mL
15	0.99913
16	0.99897
17	0.99880
18	0.99862
19	0.99843
20	0.99823
21	0.99802
22	0.99780
23	0.99756
24	0.99732
25	0.99707
26	0.99681
27	0.99654
28	0.99626

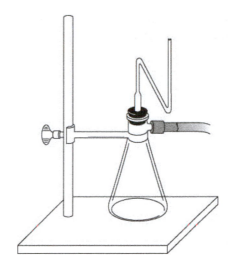

2-5. Refill the pycnometer with water and return to the same analytical balance to weigh the filled pycnometer a second time. Repeat the necessary steps until you have at least three masses for the pycnometer filled with water.

The density of water at the temperature at which you carry out the experiment must also be known (Procedure 2-2). Because the density of a liquid changes with temperature (generally the density decreases with increasing temperature), the temperature at which a density was determined must be known. The temperature is provided as a superscript to the symbol for density, for instance:

$$\rho_{H_2O}^{20} = 0.99823.$$

2-6. For the Report Form, you will need to calculate the average masses of the empty pycnometer, the filled pycnometer, and the mass of the water and the standard and percent deviation of each. You will need to express the average masses with significant figures only. (See Common Procedures and Concepts Section at the end of this manual.) Using the density for water at the temperature you recorded, you are to calculate the volume of the pycnometer. Again, express this volume with the correct number of significant figures.

Determination of the Density of an Unknown Liquid

3-1. Your instructor will provide a sample of an unknown liquid. Record the number or code for the unknown. Use a vacuum flask fitted with the one- or two-holed stopper to empty and dry the pycnometer. Fill the pycnometer with acetone and then empty it. Do these steps two more times. Following a third rinse, draw air through the pycnometer for approximately 2 minutes.

3-2. Fill the pycnometer with the unknown liquid, return to the same balance used previously, and re-tare a watch glass. Then determine the mass of the filled pycnometer to the nearest 0.0001 g. Empty the pycnometer, refill with the unknown, and reweigh it. Repeat these steps until you have three values. Calculate the density of the unknown liquid using the masses of the filled and empty pycnometer and the volume of the pycnometer. Express your final result with the correct number of significant figures.

Consult the Common Procedures and Concepts Section at the back of this manual for details on the calculation of standard and percent deviations, the determination of significant figures, etc.

NOTE: Follow your instructor's directions for disposal of materials. If you wish to and are allowed to keep your pycnometer, empty it, rinse it with acetone, and pass air through the pycnometer for 2 minutes.

NOTE: Ask your instructor if you did not understand the purpose of using the acetone in Procedure 3-1.

2   EXPERIMENT 2: DENSITY MEASUREMENTS

Prelab Exercises

1. What are the units of density? Can density have other units? If so, give examples.

2. Will the numerical value of density change with changing units? Explain.

3. What volume will 26.50 g of water at 24.0°C occupy? (Hint: See Table 2.1.)

4. How is the volume of the pycnometer determined in this experiment?

5. Define five concepts associated with this experiment.

6. Describe three techniques used during this experiment.

Date _____   **Student's Signature** _____

Name (Print)		Date (of Lab Meeting)	Instructor

Course/Section		Partner's Name (If Applicable)

2   EXPERIMENT 2: DENSITY MEASUREMENTS

Report Form

DATA

Mass of empty pycnometer: _____ g   _____ g   _____ g

Mass of pycnometer filled with water: _____ g   _____ g   _____ g

Temperature of laboratory/water: _____ °C

Density of water at this temperature: _____ g mL^{-1} (from Table 2.1)

Unknown No.: _____

Mass of pycnometer filled with unknown: _____ g   _____ g   _____ g

Date _____   **Instructor's Signature** _____

ANALYSIS

Show a sample of each calculation in your lab notebook.

Average mass of empty pycnometer: _____ g

   Standard deviation: ± _____ g

   Percent deviation: ± _____ %

Average mass of pycnometer filled with water: _____ g

   Standard deviation: ± _____ g

   Percent deviation: ± _____ %

Average mass of water: _____ g

   Percent deviation: ± _____ %

(See the discussion of Propagation of Error in the Common Procedures and Techniques Section at the end of this manual.)

Average volume of pycnometer: _____ mL

   Percent deviation: ± _____ %

Unknown No.: _____

Average mass of pycnometer filled with unknown: _____ g

  Standard deviation: ± _____ g

  Percent deviation: ± _____ %

Average density of unknown: _____ g mL^{-1}

  Percent deviation: ± _____ %

POSTLAB QUESTIONS

1. Describe three techniques used in this experiment that relate to working with glass.

2. What was the purpose of the acetone in this experiment?

3. If you were to repeat this experiment, what would you do differently to improve your results and the amount of time it took you to obtain your data?

4. If two students had different unknowns but used the same pycnometer, how would their data and results compare?

5. Was the calculated density of the unknown accurate or precise? Is it possible that the calculated density of the unknown was precise, but not accurate?

6. What was the purpose of the aspirator in this experiment?

7. Is density an intensive or extensive property? Explain why.

8. Answer any questions assigned by your instructor.

Date _____ **Student's Signature** _____

Cost of a Chemical Product

A Guided Inquiry Experiment

INTRODUCTION

How is the price you pay for a product in the store derived? Do the prices often seem outrageously high? Have you questioned why a product sells for several dollars when the raw materials from which it is derived are worth only a few pennies?

When comparing the retail cost of a product with the cost of materials, consumers may forget to consider the various costs for converting the raw materials to products. Manufacturers must pay for labor, energy, machinery, buildings, taxes, environmentally acceptable waste disposal, transporting the products to the markets, and advertising. They must allow for a reasonable profit. Industry relies on mass production and efficient machinery to keep the cost per item low and the profit high. Most industries deal with very high volumes or they diversify.

This experiment will acquaint you with the many "hidden" costs of a product. Be prepared for your product to be alarmingly expensive, since you will not have the advantages of either large-scale production or diversification.

OBJECTIVES

In this experiment, you will investigate the effect of heating a hydrated salt and use the mass change to propose a chemical equation for the changes taking place during heating. You will act as the operator of a chemical plant (on a laboratory scale) whose product is formed by heating Epsom Salt.

CONCEPTS

This experiment uses the concepts of mass, economics, and stoichiometry. You will test vapors produced for the presence of hydrogen gas, oxygen gas, or water vapors. You will control a reaction and make quantitative transfers.

TECHNIQUES

Lighting and adjusting a burner, weighing on the top loader balance, writing equations, and working with hot objects are used.

ACTIVITIES

You, someone in your class, or the Chemistry Department will buy the Epsom Salt that you will need at a grocery or drug store. You will set up the equipment in the laboratory to heat the raw material. You will need expert advice on how to run the process efficiently; you will need to have your operation inspected to make sure your plant is in compliance with local, state, and federal regulations; and you will have to package your product and keep track of all your expenses. You will calculate a price for your product that allows for a profit. You then negotiate with your instructor a fictitious price for your packaged product and calculate your profits or losses.

CAUTION

Use approved eye protection. Be aware of chemicals and hot equipment. Follow the details provided by your instructor and by the engineer's suggested plant design. Remember that glass stirring rods are fragile.

PROCEDURES

Plant Personnel

1-1. At various times during this experiment you are the different "personnel" (scientist, chemist, engineer, plant operator, worker, salesman, and business manager). You must keep a record of the time spent on each task to the nearest one-quarter hour. Pay yourself $12.00 per hour for the time spent in each of the tasks.

The Safety Inspector

2-1. Your instructor will play the role of government safety inspector. He or she will inspect your setup, see that all general safety rules are followed, and check to insure that all cautions are observed. Fines may be imposed that will result in lowered grades.

The Scientist

3-1. You are the research scientist for the chemical company. You are to investigate the reaction(s) that occur when Epsom Salt is heated. In the literature, you have found that Epsom Salt is magnesium sulfate heptahydrate ($MgSO_4 \cdot 7H_2O$). The crystals of the heptahydrated salt are reported to be transparent. However, the literature also states that at room temperature and low humidity the heptahydrated salt will readily lose one molecule of water. The result would be the hexahydrated compound. The hexahydrated material formed in this way is reported to consist of dull, white, highly fractured crystals. Knowing these facts, you can visually determine which of the two forms of Epsom Salt is the major constituent in your starting material.

3-2. You have been told that heating Epsom Salt yields a product that can be used as a drying agent. Your task is to determine what is being driven off during the heating process.

Place Epsom Salt to a depth of about 1 cm in an 18 × 150 mm test tube. Place a 250-mL beaker on the top-loader balance and tare the balance and beaker. Put the test tube with Epsom Salt into the beaker and record the mass of the test tube and Epsom Salt.

3-3. Use a test tube holder to hold the test tube with Epsom Salt. Gently heat the closed end of the test tube with the Bunsen burner.

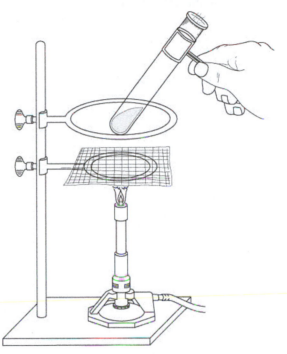

CAUTION

Make sure that you do not point the open end of the test tube toward other people in the lab. Uneven heating can cause materials to be propelled from the mouth of the test tube.

Heat the test tube for 3–5 minutes. Look for changes during heating, gases given off, condensation near the mouth of the test tube, etc. Determine the identity of any gases produced. If a gas is given off at this stage, it will be a single compound: water, hydrogen, or oxygen (see Common Procedures and Concepts Section at the end of this manual). Perform tests needed to identify any gas. Consult with your safety inspector concerning procedures.

3-4. Allow the test tube and its contents to cool. Place the beaker back on the balance and tare the balance. Place the test tube in the beaker and determine the mass of the test tube and heated Epsom Salt. Record your findings.

Purchasing Department and Plant

4-1. Prior to the laboratory period, someone in your lab section or in the stockroom must go to a store and purchase a one-pound carton of Epsom Salt. You will need only about 20 g of this salt. A sales receipt showing this purchase must be kept as a record and brought to the laboratory. You must also keep a record of the time spent on your trips to the store to the nearest one-quarter hour. (If the Epsom Salt is furnished, your instructor will provide you with a value to be used as the time spent obtaining your sample and the price paid.)

4-2. Your laboratory fee is considered your capital investment in the plant and equipment. It gives you access to the laboratory and standard equipment.

The Engineer

5-1. As an engineer, you are to design your experiment in compliance with the following knowledge. Heating $MgSO_4 \cdot 7H_2O$ or $MgSO_4 \cdot 6H_2O$ should be carried out in a 250-mL beaker. A 250-mL beaker is a good choice because it is of adequate size, is relatively inexpensive, and can withstand high temperatures. Extra precautions must be taken to prevent breakage of the glass beaker for the safety of the operator. A ringstand and iron ring are needed to secure the beaker above the burner. Never have the beaker in direct contact with the hot iron ring. Place a wire gauze between the beaker and the lower iron ring. Stirring the material in the beaker during the heating process may cause the beaker to work its way off the gauze, fall, and break. To prevent such a destructive occurrence and the subsequent loss of time and material, a second iron ring must be placed around the beaker. This ring does not have to touch the beaker. (This concludes the work of the engineer. Record the time that you have spent as the engineer.)

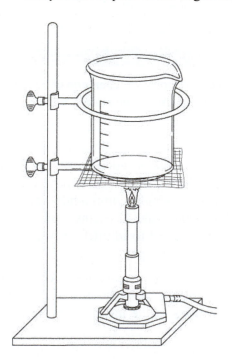

The Plant Operator

6-1. You have progressed to plant operator. Take your beaker to a top-loader balance. Tare the balance with the beaker on it. Place approximately 20 g of Epsom Salt in the beaker. (The amount can vary from 19 to 21 g.) Determine the mass of your Epsom Salt sample. Record this mass in your notebook. Also record your time.

6-2. Put the beaker with the Epsom Salt on the ringstand. Make sure that the beaker is secure and all the equipment is ready.

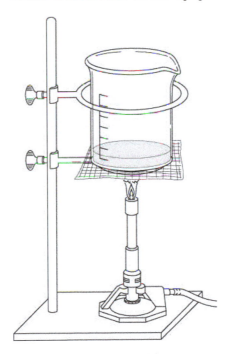

The Safety Inspector

7-1. Have your instructor, in his/her role as government safety inspector, inspect your plant to verify that you are complying with all regulations. He/she will deduct points as a "fine" if you are found to be without eye protection, have left the balances without having cleaned up any spillages, or are caught practicing other unsafe laboratory practices. Severe violations could result in closure of your plant.

The Plant Operator

8-1. When your plant has been approved, record the time you light the burner, and then gently and cautiously heat the beaker containing the Epsom Salt, making sure that your product does not melt. Reduce the size of the flame on the burner if any melting is observed. Use a glass rod to stir the crystals to allow the reaction to occur.

NOTE: If you heat too strongly and do not stir the crystals sufficiently, the reaction will occur too quickly. Too much heat too early in the operation will cause your material to cake. If this happens, your final product will not be the desired white powder and the equipment will be difficult to clean. Since some of the material will stick to the walls and will be unrecoverable, the yield will be low.

After you have **gently** heated the material for a few minutes without any clumping occurring, slowly increase the heating while making sure that your product does not begin to melt. Continue gently heating and stirring until you have a fine, white powder.

8-2. After completion of the reaction, allow it to cool to the touch. Record the time you turned off the burner. At the top-loader balance, tare the balance and weigh the beaker and sample. Take the beaker back to the lab bench and heat until you observe vapors given off. Continue to heat for an additional 2 minutes but not more than 4 minutes total. Repeat the cooling and weighing. If the two mass values are within 0.10 g, go to the next step. If not, repeat the heating, cooling, and weighing until two consecutive weighings are within 0.10 g of each other or until your instructor approves your stopping. Too much heat will cause the product to be an unattractive gray color and will pollute the air with sulfur trioxide. Polluting the air will result in your plant being closed.

Tare the top-loading balance with a piece of paper on it. Transfer the sample from the beaker to the paper, and then weigh the product. Calculate the percent yield based upon the starting material previously determined in Procedure 6-1 above.

8-3. Your plant operation may generate by-products which "pollute" the environment. Leave a pleasant, unpolluted environment for the next generation (the students who use the lab after you). You must clean all the glassware you used, clean your working station, and clean around the balances. To clean the beaker, add 50 to 75 mL of water and heat the beaker until the residue is released from the glass walls of the beaker. Then thoroughly clean the beaker and stirring rod. Do not use the stirring rod as a tool for cleaning the beaker.

You will be assessed a fine in the form of points deducted from your grade if there is any evidence of pollution.

The Business Manager

9-1. Use the raw material costs, equipment costs, labor costs, energy costs, and a reasonable profit to calculate a price per gram of your product. The data in the Report Form can serve as a guide for this calculation. Calculate your profit or loss. In your notebook record all data needed to complete the Report Form.

The Sales Department

10-1. Submit your labeled, packaged product (with decorated packaging if you like) to your instructor, who is now playing the role of the purchaser. The package must show the mass of your material. Use your best sales pitch to negotiate with your instructor for a price for your packaged product. Your instructor may bargain with you on the yield and quality of your product. Be prepared to promote your product. One possible selling point is its possible use as a valuable drying agent. Your instructor is also going to inspect your work area to ensure that it has been left clean and that all equipment is returned to its proper location. Ask your instructor to sign your notebook and Report Form.

Name (Print) | Date (of Lab Meeting) | Instructor

Course/Section

3 **EXPERIMENT 3: COST OF A CHEMICAL PRODUCT**

Prelab Exercises

1. What is the purpose of this experiment?

2. How many waters of hydration are present in Epsom Salt? Give the chemical formula for Epsom Salt.

3. What are the gas possibilities from Epsom Salt, based on its chemical formula, if you ignore any sulfur containing gases?

4. In your own words, describe the following:

hexahydrate

taring a balance

adjusting the air vent on a burner

5. Define five concepts associated with this experiment.

6. Describe two techniques used to prevent accidents during this experiment.

7. List the **three** roles that your instructor will play in this experiment.

Date _____ **Student's Signature** _____

3 EXPERIMENT 3: COST OF A CHEMICAL PRODUCT

Report Form

DATA

A. Test on the Heating of Epsom Salt

Mass of Epsom Salt and test tube before heating: _____ g

Mass of Epsom Salt and test tube after heating: _____ g

Appearance of starting material:

Observations (tests) of vapors produced when Epsom Salt is heated:

B. Chemical Plant

Raw Material

Cost of purchased package of Epsom Salt: $ _____ (attach receipt, if available)

Mass of Epsom Salt in purchased package: _____ g

Chemical Data

Mass of Epsom Salt used: _____ g   Mass of product obtained: _____ g

Energy

Time burner lit: _____ min.   Time burner turned off: _____ min.

Time burner re-lit: _____ min.   Time burner turned off 2nd time: _____ min.

Time burner re-lit: _____ min.   Time burner turned off 3rd time: _____ min.

Personnel (round time to nearest ¼ hour; minimum ¼ hour)

Scientist (developing the process):   _____ hr

Purchasing Department (trip to store):   _____ hr

Engineer (building of setup):   _____ hr

Plant Operator (run the process):   _____ hr

Plant Operator (cleanup): _____ hr

Business Manager (packaging the product
and calculation of price): _____ hr

Consultant (help from instructor): _____ hr

Safety Inspector: _____ hr

Salesman (bargaining with instructor): _____ hr

C. Quality of Product (Instructor's Evaluation):

☐ Reagent grade   ☐ Tech grade   ☐ Poor   ☐ Unsatisfactory

D. Government Safety Inspection:

☐ In compliance with all rules and regulations

☐ Penalty for infringement (number of grade points to be deducted)

Describe your product, its uses, how it was packaged, etc.

Date _____   **Instructor's Signature** _____

ANALYSIS

A. Tests Results

Mass lost upon heating of the test tube and Epsom Salt sample: _____ g

Identity of vapors produced: _____

B. Proposed Equation for the Change Taking Place

Mass of Epsom Salt: _____ g

Mass of product: _____ g

Mass change: _____ g

Formula of your Epsom Salt based on its appearance (circle one):

$MgSO_4 \cdot 7H_2O$ or $MgSO_4 \cdot 6H_2O$

Moles of Epsom Salt: _____ mol   Moles of vapor given off: _____ mol

Possible equation for the reaction in your experiment based on the amounts above and the identities of the starting material and the vapors. Explain how you determined the coefficients and products:

Percent yield (based upon the equation above): _____ % (See Postlab Question 3 below).

C. Costs

Energy

Total time that the burner was lit: _____ min.

Cost of gas (min × $0.01/min): $ _____

Capital Investment

$ _____ (10% of laboratory fee paid but not less than $6.50)

Personnel (round time to nearest ¼ hour; minimum ¼ hour)

Scientist (developing the process): _____ hr × $12.00/hr = $ _____

Purchasing Department (trip to store): _____ hr × $12.00/hr = $ _____

Engineer (building of setup): _____ hr × $12.00/hr = $ _____

Plant Operator (run the process): _____ hr × $12.00/hr = $ _____

Plant Operator (cleanup): _____ hr × $12.00/hr = $ _____

Business Manager (packaging and pricing): _____ hr × $12.00/hr = $ _____

Consultant (help from instructor): _____ hr × $12.00/hr = $ _____

Safety Inspector: _____ hr × $12.00/hr = $ _____

Salesman (bargaining with instructor): _____ hr × $12.00/hr = $ _____

Total Personnel Expenses: = $ _____

Total Costs

Raw material: $ _____

Energy: $ _____   Sales price obtained: $ _____

Capital investment: $ _____   Profit or loss before taxes: $ _____

Personnel: $ _____   Total cost to produce 1.0 g of product: $ _____

Total costs: $ _____

D. Profit

$$\text{Percent profit after taxes} = 100 \times \frac{\text{Profit before taxes} - 27\% \text{ of profit before taxes}}{\text{Sales price of product}} = \text{_____} \%$$

POSTLAB QUESTIONS

1. Draw a particle representation of the chemical reaction that took place in your chemical plant.

2. Discuss the role of the research scientist.

3. If your percent yield was under 100%, what does this mean and how would the plant operator react to this information? How would the plant operator explain percent yield that is over 100%?

4. If Epsom Salt is heated too much, a gray product is formed and sulfur trioxide gas is given off. Write an equation and draw a particle view of the reactants and products.

5. How does heated Epsom Salt act as a drying agent?

6. Answer any questions assigned by your instructor.

Date _____ **Student's Signature** _____

Soap Making

A Skill Building Experiment

INTRODUCTION

The manufacture and use of soap has been known for nearly as far back as written records. The Phoenicians made and used soap before 600 BC; soap was used by the Roman Empire, including Gaius Julius Caesar. Soap was originally made by boiling animal fat with extracts of wood ashes. Eventually, lye (sodium or potassium hydroxide) replaced the wood ashes. The product known as soap is the sodium or potassium salt of the fatty acids found in animal fats or vegetable oils. After World War II, synthetic cleaning agents, known as detergents, displaced soap as the most popular cleaning agent. Detergents are made from petroleum-derived hydrocarbons.

This experiment will acquaint you with the way soap was made from fats and oils in most households a century ago. Since we can quickly and easily purchase any number of cleaning agents, making soap is not an activity of most households but is often part of historic reenactment sites. The skill of making soap has experienced a resurgence in popularity as many farmers' markets and other distributors now carry homemade soaps that feature purity and fats or oils from various sources (goat's milk, coconut, avocado, etc.).

Animal fats and vegetable oils contain compounds called glyceryl esters. Many glyceryl esters found in nature contain the carboxylate (fatty acid) component of stearate (see the equation below). Other glyceryl esters contain more than one kind of carboxylate (fatty acid) component (the 16 in the $(CH_2)_{16}$ will vary). When any of these compounds (esters) are heated with a solution of a base such as sodium hydroxide in water, the esters are hydrolyzed into glycerol (the alcohol component) and salts of fatty acids (the carboxylic acid component). This process, known as saponification, is described by the following equation in which the tristearate is used as an example. For our purposes here, assume that you are using glyceryl tristearate.

glyceryl tristearate   glycerol

Glycerol is soluble in water. The long-chain sodium carboxylates (soap) are less soluble in water than glycerol and can easily be separated from other compounds in the solution by adding a concentrated sodium chloride solution to precipitate the soap.

Common fats and oils may contain different carboxylic acids. Many soaps also contain the potassium salts of the acids. Some examples of carboxylic acids found in fats and oils follow.

lauric acid $CH_3(CH_2)_{10}CO_2H$

linoleic acid $CH_3(CH_2)_4CH{=}CHCH_2CH{=}CH(CH_2)_7CO_2H$

oleic acid $CH_3(CH_2)_7CH{=}CH(CH_2)_7CO_2H$

palmitic acid $CH_3(CH_2)_{14}CO_2H$

stearic acid $CH_3(CH_2)_{16}CO_2H$

Above are the condensed formulas for several large carboxylic acids. The line bond formula for lauric acid would be:

$$H-\overset{\overset{\displaystyle H}{|}}{\underset{\underset{\displaystyle H}{|}}{C}}-\overset{\overset{\displaystyle H}{|}}{\underset{\underset{\displaystyle H}{|}}{C}}-\overset{\overset{\displaystyle H}{|}}{\underset{\underset{\displaystyle H}{|}}{C}}-\overset{\overset{\displaystyle H}{|}}{\underset{\underset{\displaystyle H}{|}}{C}}-\overset{\overset{\displaystyle H}{|}}{\underset{\underset{\displaystyle H}{|}}{C}}-\overset{\overset{\displaystyle H}{|}}{\underset{\underset{\displaystyle H}{|}}{C}}-\overset{\overset{\displaystyle H}{|}}{\underset{\underset{\displaystyle H}{|}}{C}}-\overset{\overset{\displaystyle H}{|}}{\underset{\underset{\displaystyle H}{|}}{C}}-\overset{\overset{\displaystyle H}{|}}{\underset{\underset{\displaystyle H}{|}}{C}}-\overset{\overset{\displaystyle H}{|}}{\underset{\underset{\displaystyle H}{|}}{C}}-\overset{\overset{\displaystyle H}{|}}{\underset{\underset{\displaystyle H}{|}}{C}}-\overset{\overset{\displaystyle O}{\|}}{C}-O-H$$

OBJECTIVES

During this experiment, you will gain experience in working with solutions, stoichiometry, the triple-beam balance, filtration, and materials that are difficult to transfer. You will also observe some physical and chemical properties.

CONCEPTS

This experiment uses the following concepts: mass, volume, solution preparation, physical and chemical properties, and stoichiometry.

TECHNIQUES

Lighting and adjusting a burner, weighing on the triple-beam balance, transferring solids and liquids, and using vacuum filtration are techniques used in this experiment.

ACTIVITIES

You will convert a sample of fat or vegetable oil into soap (saponification). You will isolate your product, purify it, and describe it. You will need to discuss physical and chemical properties of the soap that you make in this experiment. You will also be asked to discuss the molecular process by which the soap was formed. (In an organic course, you would go into even greater detail on the changes that take place at the molecular level.)

> **CAUTION**
>
> **You will be working with flames and flammable materials. (It is better to prevent a fire than to have to react to one.) You will also be working with potentially caustic solutions. Avoid direct contact with any solution that is hot or contains sodium hydroxide. Wear approved safety goggles at all times.**

PROCEDURES

From Oil or Fat to Soap

1-1. Use the triple-beam or top-loader balance to weigh a clean, dry 400-mL beaker to the nearest 0.1 g. Put about 10 g of oil or fat in the beaker. A range of 8.5 to 11.5 g of oil or fat in the beaker is permissible. Record the actual mass of beaker and sample to the nearest 0.1 g. Use a spatula to transfer the material if you are using a semisolid fat or shortening. If an oil is used, pour the liquid down a stirring rod or use the fact that 10 g will have a volume of about 14 mL.

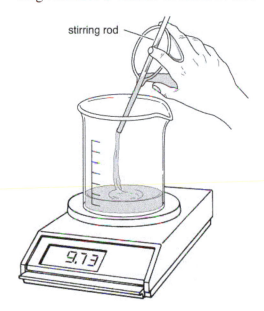

stirring rod

1-2. Use the triple-beam or top-loader balance to weigh 20 g of solid sodium chloride (salt) into a 250-mL Erlenmeyer flask. Record the mass of the flask and of the flask plus salt to the nearest gram. Add approximately 50 mL of distilled water to the salt, stopper the flask, and shake the stoppered flask. Periodically shake the flask during the following procedures or until most of the sodium chloride has dissolved. The solution needs to be a saturated solution before it is used in a later step. If the solution is cloudy, filter it by vacuum filtration.

1-3. Use your graduated cylinder to obtain 20 mL of 95% ethanol from its storage bottle. Pour the ethanol into the beaker containing the oil (or fat). Carefully stir the mixture with a glass stirring rod until the mixture is nearly homogeneous. Use the graduated cylinder used above (it does not need to be dry) to obtain 25 mL of 10% (~2.5 *M*) aqueous sodium hydroxide solution. Record in your notebook the actual volume and molarity of the NaOH solution used. Add the sodium hydroxide solution to the beaker containing the oil (or fat) and ethanol. On the outside of the beaker, use a grease pencil to mark the liquid level of the solution present.

CAUTION	

Place a wet towel or wet sponge, large enough to cover your beaker, in a readily accessible location. This will be used should your solution catch fire in the next few steps.

1-4. Place the beaker containing the mixture of fat (or oil), ethanol, and sodium hydroxide on a wire gauze supported on a ring clamp above a laboratory burner. Use a second ring clamp around the beaker about two inches above the wire gauze to prevent the beaker from sliding off the wire gauze. Place the apparatus under the hood (at your work space, if possible). Stir the mixture until it is nearly homogeneous. Light and adjust the burner for a small flame that will not deposit carbon on the beaker. Heat and stir the mixture gently. The mixture must be warmed to near boiling and kept at near that temperature. Don't let it boil vigorously. If foam rises in the beaker, move the burner away. Continue simultaneously heating and stirring the mixture. Stirring is necessary for mixing the reagents, permitting the reaction to proceed more quickly, and preventing dangerous over-heating and bumping.

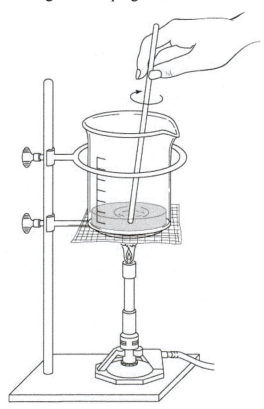

WARNING

Keep the flame away from the top of the beaker where ethanol vapors might easily ignite. In case of fire, turn off your burner at the bench gas valve and place the wet towel or wet sponge on top of the burning beaker. Keep the beaker covered until the flames are extinguished. Notify your instructor. You should be able to continue after replacing part of the ethanol lost during the fire.

1-5. Keep stirring and maintain the heating at or near a gentle boil for at least 20 minutes. Rinse particles clinging on the inside of the beaker back into the reaction mixture by using your squeeze bottle to squirt very small amounts of distilled water around the inside of the beaker. Allow the liquid level inside the beaker to gradually decrease to about half its original volume. Foaming commonly occurs when the reaction nears completion.

After about 20 minutes of gentle heating, your beaker should contain an off-white to yellowish viscous mass (a mixture of soap, glycerol, and excess sodium hydroxide solution). (If the volume is more than about 50% of the original volume, allow additional liquid to evaporate by additional boiling.) Turn off the burner. Let the beaker cool for 3 to 4 minutes, while you prepare an ice bath from a 600 mL or larger beaker half-filled with ice and tap water. Using a folded towel, beaker tongs, or hot mitt, remove the beaker containing the soap mixture carefully from the ring support and set it into the ice bath. Let the mixture cool in the ice water to below room temperature. Replenish the ice as necessary.

1-6. Remove the beaker from the ice bath. Being careful not to transfer any of the undissolved sodium chloride, pour the saturated sodium chloride solution into the beaker containing the now cool reaction mixture. Stir the mixture thoroughly but gently. The excess sodium hydroxide and the glycerol will remain dissolved; the soap should be precipitated. The reaction mixture must be mixed thoroughly with the sodium chloride solution to obtain pure soap. Use your spatula to press the soap against the beaker wall and expose some of the solution trapped in the chunks of soap floating in the mixture. Improper execution of this step will leave sodium hydroxide and glycerol trapped in the soap and the product will require aging before it is safe for one to touch.

NOTE: During the aging process any NaOH present will react with carbon dioxide from the air and produce sodium bicarbonate ($NaHCO_3$), which is much less caustic.

1-7. Set up the vacuum filtration equipment. (See the Common Procedures and Concepts Section at the end of this manual.) Line the funnel with a filter paper that has been moistened with distilled water. Turn on the aspirator. Decant the supernatant liquid through the funnel. Transfer the soap to the funnel and draw air through it to remove some of the water retained by the soap. Disconnect the hose. Turn the aspirator off.

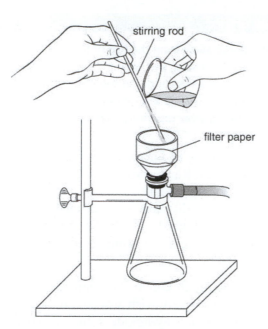

stirring rod

filter paper

1-8. Add 50 mL of distilled water to a 100-mL beaker and allow the beaker to chill in the ice bath. (You may also chill your distilled water bottle in the ice bath.) With the aspirator turned off, use about 25 mL or half of the cold water (no ice) to rinse any remaining soap from the beaker into the funnel. Scrape soap from the inside of the beaker with your spatula. Also use your spatula to stir the soap/ice-water mixture in the funnel into a slurry. Do not tear the filter paper. Turn on the aspirator to filter the mixture.

1-9. Prepare a hot water bath. Check for residual sodium hydroxide in your soap by placing a sample of the soap approximately half the size of a pea into a small beaker or test tube and add 4 to 6 mL of 95% ethanol. Place the test tube or small beaker into the hot water bath. Gently stir the mixture until the soap has dissolved. Cool the solution to room temperature before adding two or three drops of phenolphthalein indicator solution. If the solution turns bright pink, there is still a considerable amount of sodium hydroxide trapped in the soap. If so, the sodium hydroxide will react with carbon dioxide in the air as you allow the soap to age for two or more days.

1-10. Weigh a clean piece of notebook paper on the triple-beam or top-loader balance. Put your soap in the center of this sheet of paper and discard the filter paper. Wrap the notebook paper around the soap making a neat package. Weigh this soap and package before showing it to your instructor. Ask your instructor to rate your product as either "technical grade" or "reagent grade" and to initial your Report Form where it asks for "Instructor's initials."

1-11. Calculate the mass of soap obtained.

Some Chemical and Physical Properties of Soap

2-1. Use some of the soap you just made (or some furnished by your instructor) to wash your hands. Soap should not feel greasy but the soap you prepared might.

2-2. Rub some chalk dust on your hands. Use soap and water to rinse your chalky hands. Observe carefully. Does the soap produce lather? What does it take to get your hands clean?

2-3. Describe at least three physical and three chemical properties that you could verify using your soap sample and materials in the lab. Carry out these tests as time and materials allow.

Clean Up

3-1. All materials should be disposed of as your instructor directs. Clean all glassware and return it to its proper place. Have your instructor sign your notebook and Report Form.

EXPERIMENT 4: SOAP MAKING

Prelab Exercises

1. Below is an equation for the production of soap. This equation is written using condensed formulas for sodium stearate and glycerol. The formula of every reactant and product except the glyceryl tristearate is shown. Provide the missing condensed formula for glyceryl tristearate. (A condensed formula has all the atoms of a given element combined.) Record this equation in your notebook. This equation will be used for the stoichiometric calculations in this experiment.

$$\underline{\hspace{4cm}}\text{(s)} + 3\text{NaOH(aq)} \rightarrow 3C_{18}H_{35}O_2Na\text{(s)} + C_3H_8O_3\text{(}l\text{)}$$

   glyceryl tristearate   sodium hydroxide sodium stearate   glycerol
    (fat or oil)   (soap)

2. What is the purpose of this experiment?

3. Describe the equipment used during vacuum filtration. Explain the main precaution you should take when using vacuum filtration.

4. What are physical properties? List some physical properties.

5. What are chemical properties? List some chemical properties.

6. What is saponification?

7. In the Introduction, the line bond for glycerol is given. Give a condensed formula for glycerol. Then give the line bond formula for acetic acid.

Date _____ **Student's Signature** _____

Name (Print) _____   Date (of Lab Meeting) _____   Instructor _____

Course/Section _____   Partner's Name (If Applicable) _____

4 EXPERIMENT 4: SOAP MAKING

Report Form

DATA

Mass of beaker: _____ g   Volume of 95% ethanol: _____ mL

Mass of beaker plus fat or oil: _____ g   Volume of ~2.5 M NaOH: _____ mL

Mass of fat or oil: _____ g   Molarity of NaOH: _____ M

Mass of soap obtained: _____ g

Technical grade ☐   Reagent grade ☐   Instructor's initials _____

Describe the procedures you used to verify physical properties of your soap and describe the results obtained.

Describe the procedures you used to verify chemical properties of your soap and describe the results obtained.

Date _____   **Instructor's Signature** _____

ANALYSIS

(Show all calculations below each quantity.)

Moles of glyceryl tristearate (fat or oil) used: _____ moles

Moles of NaOH used: _____ moles

Moles of sodium stearate (soap) produced: _____ moles

Which reactant is the limiting reactant? _____ Explain your choice.

Theoretical yield of soap based upon _____ being the limiting reagent is:

_____ moles of soap or _____ g of soap

Percent yield: _____%

List some physical properties of soap:

List some chemical properties of soap:

POSTLAB QUESTIONS

1. Describe techniques used in this experiment that relate to working with hot, caustic solutions. (Hint: Look at construction of the equipment and the caution statements.)

2. What is taking place as soap made by this method is allowed to age?

3. What was the purpose of phenolphthalein in this experiment?

4. What is the purpose of the ethanol in this experiment?

5. Explain why the soap you made feels greasy.

6. What changes do you propose to increase your percent yield?

Date _____   **Student's Signature** _____

Reactions of Calcium

A Guided Inquiry Experiment

INTRODUCTION

Calcium is an alkaline earth metal. Its ionic form (Ca^{2+}) is needed by our bodies for normal growth and development. Calcium metal and calcium ions have very different physical and chemical properties. One such difference is illustrated when each is combined with water or acids. For example, the combination of calcium metal and water in this experiment does not mirror what happens when calcium ions like those present in food are combined with water.

OBJECTIVES

You will investigate the reactions of calcium with water and with hydrochloric acid. You are to then propose equations for the reactions observed and gather mass data to support or refute these equations.

CONCEPTS

This experiment uses the concepts of mass and stoichiometry.

TECHNIQUES

Reading a buret, graphing experimental data, transferring liquids and solids, and identifying hydrogen, oxygen, water, or chlorine vapors are some of the techniques utilized.

ACTIVITIES

In this experiment, you are to collect data concerning reactions of calcium with water and with HCl. Like many research chemists, you are to propose an equation that fits your observations, and then collect data to see if there is support for your proposed equation. You will be asked to collect data and to analyze that data for patterns, much as research scientists do. The point is that you are looking for meaning in your data.

CAUTION

Avoid contact with acids or bases. Flush with water in case of exposure. Wear approved eye protection. Dispose of materials as directed by your instructor.

PROCEDURES

Combining Calcium and Water

1-1. Fill a 250-mL beaker ½ full of water. Obtain 1 small piece or a small amount of calcium. Observe the appearance of calcium metal. Record your conclusions.

1-2. Fill a test tube with water, cover the mouth with your finger, and invert the filled test tube into the beaker. Remove your finger when the mouth of the test tube is below the water level in the beaker. Drop in the small piece of calcium into the beaker. Move the test tube so that it is directly over the piece of calcium. Record your observations.

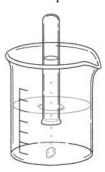

1-3. Test the resulting solution with litmus paper. Blue litmus paper turning to red indicates the presence of H^+ (acid). Red litmus turning to blue indicates the presence of OH^- (base). The color of litmus paper indicates whether it was last in contact with H^+ or OH^-.

1-4. Test any gas produced to determine if it is hydrogen, oxygen, or water (see Common Procedures and Concepts Section at the end of this manual). Collect more gas if needed to complete your tests. Record your observations.

1-5. Discard the contents of the test tube as directed by your instructor and clean all glassware used thus far.

Measuring Mass of Calcium and Hydrochloric Acid

2-1. Obtain two small test tubes, one 100-mL beaker, and two droppers or disposable pipettes. Put distilled water in one of the test tubes. Label one dropper with an "A" and the other with a "B." Find the number of drops per 1 mL for each of your droppers. This can be done by counting the drops required to fill from 2.0 to 3.0 mL of a 10-mL graduated cylinder. Be sure to read the bottom of the meniscus.

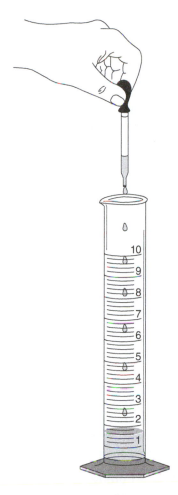

2-2. Carefully rinse and dry two small test tubes. Your instructor has set up two burets to dispense solutions. Label one test tube "HCl" and obtain about 7 mL from the buret containing 1.__ __ M HCl. Label the other test tube "NaOH" and place about 7 mL of 1.__ __ M NaOH in it from that buret. In both cases, record the EXACT molarity to three significant figures. Use the dropper labeled "A" for the acid solution and the one labeled "B" for the base solution. Calculate the number of moles per drop of the HCl and NaOH.

2-3. Clean and dry a 100-mL beaker. Place it on the analytical balance and tare the balance. Transfer into the beaker between 0.062_ and 0.078_ g of calcium turnings. Record the mass of the turnings transferred.

2-4. Add the 1.__ __ M HCl dropwise to the beaker containing calcium until you have added 5 mL. Swirl the beaker until all evidence of a reaction has ceased. Calculate the number of moles of HCl that you have added.

2-5. Add 1 drop of methyl orange to the beaker. A reddish orange color indicates the presence of acid. Yellow indicates a more neutral solution. You may need to put a white paper under your beaker to clearly distinguish the color. When the reaction of calcium and hydrochloric acid ceased, what reactant was in excess?

2-6. Add 1._ _ *M* NaOH dropwise to the beaker, counting and swirling after each drop. Record the number of drops needed to cause the color to change to yellow. Calculate the number of moles of NaOH required to react with the excess reactant.

2-7. Repeat the above procedures for two additional runs. Put your data on the board and copy data from other groups in your class.

Clean Up

3-1. All materials should be disposed of as your instructor directs. Clean all glassware and return it to its proper place. Have your instructor sign your notebook and Report Form.

5   **EXPERIMENT 5:   REACTIONS OF CALCIUM**

Prelab Exercises

1. What is the purpose of this experiment?

2. Describe the procedures involved in testing a gas to identify it as:

   a. water

   b. oxygen

   c. hydrogen

3. Describe the characteristics of chlorine gas.

4. Why was it important to use dropper A for HCl in the experiment and not pick a new dropper to use?

5. When determining the number of drops per mL, why is it important to hold the dropper in the vertical position?

Date _____   **Student's Signature** _____

| 5 | EXPERIMENT 5: REACTIONS OF CALCIUM |

Report Form

DATA

Combining Calcium and Water

Observations of the calcium.

Observations of the combination of calcium and water.

Results from litmus and gas tests

Measuring Mass of Calcium and Hydrochloric Acid

Dropper #1 = _____ drop/mL   Dropper #2 = _____ drop/mL

_____ M HCl   _____ M NaOH

_____ mol/drop HCl   _____ mol/drop NaOH

Drops of NaOH added after the 5 mL of HCl were added:

_____ Trial #1   _____ Trial #2   _____ Trial #3

	mass Ca	mol HCl Added	mol NaOH	mol Ca	mol HCl Excess After Ca Reaction	mol HCl Used in Ca Reaction
Trial #1						
Trial #2						
Trial #3						
Class data						
Class data						
Class data						
Class data						
Class data						
Class data						
Class data						
Class data						
Class data						
Class data						
Class data						
Class data						
Class data						
Class data						
Class data						
Class data						

(As a minimum, the first three columns must be completed to share.)

Date _____   **Instructor's Signature** _____

ANALYSIS

Combining Calcium and Water

1. Was the combination of water and calcium a physical change or a chemical reaction? Explain your reasoning.

2. Write a PROPOSED balanced chemical equation for any reaction you observed. Explain why you chose the particular products you did.

3. Draw a particle view of the reactant and product particles in two of the reactions you observed.

Measuring Mass of Calcium and Hydrochloric Acid

1. Propose a balanced equation for the reaction of calcium and hydrochloric acid. Write the balanced equation.

2. Which reactant was in excess? Give the rationale for your answer.

3. Propose a balanced equation for the reaction of the excess reactant and NaOH. Write the balanced equation.

4. What relationships do you expect in the data?

5. What pattern do you see in the data for the reaction of Ca and HCl? Try graphing the data to obtain an algebraic relationship. Attach graph(s) to this report.

6. Does the pattern in #5 support the balanced equation that you proposed in #1 above? Explain why or why not.

POSTLAB QUESTIONS

1. Explain why you were asked to graph class data instead of just your own data.

2. Based upon your observations during this experiment which of the following reactions is likely to occur? Explain.

   a. Calcium metal plus hydrochloric acid yields calcium ion plus hydrogen gas plus chloride ion plus hydroxide ion.

   b. Calcium metal plus hydrochloric acid yields calcium ion plus hydrogen ion plus chlorine.

   c. Calcium metal plus hydrochloric acid yields calcium ion plus hydrogen gas plus chloride ion.

3. If a different alkaline earth metal was used instead of calcium, would the relationship between the new alkaline earth metal and HCl be the same as it is for calcium and HCl? Explain your answer.

Date _____ **Student's Signature** _____

Aluminum and Copper(II) Chloride

A Guided Inquiry Experiment

INTRODUCTION

Some substances that are combined form a physical mixture, like combining table salt and sand. Mixtures can be described by the percentage of each substance. Alternatively, some combined substances undergo a chemical reaction. Balanced chemical equations show the stoichiometric ratios for the moles of reactants that undergo a chemical reaction to form the moles of products. What happens when the amount of the reactants required in the balanced equation is not followed? You will work with the various combinations of aluminum and a solution of copper(II) chloride dihydrate. Solutions that contain the Cu^{2+} ion are light blue. Aluminum is a common metal.

OBJECTIVES

In this experiment, you will investigate the combination of aluminum and copper(II) chloride with varying amounts of aluminum.

CONCEPTS

This experiment uses the concepts of moles, molarity, stoichiometry, and balanced equations.

TECHNIQUES

Correct use of balances, graduated cylinders, suction filtration, and hot water baths are some of the techniques encountered in the experiment.

ACTIVITIES

You will observe the outcome of combining varying amounts of aluminum with a set amount of copper(II) chloride solution. Next you will react a set amount of aluminum with a copper(II) chloride solution that has an unknown molarity. You will use these reactions to draw generalities.

PROCEDURES

1-1. Set up a hot water bath in a 250-mL beaker that is ⅔ full of water. Bring the water to a gentle boil using a hot plate while you go on to Procedure 1-2, while the water is heating.

1-2. Obtain five clean and dry 18 × 150-mm test tubes. Number them as 1 to 5.

1-3. Select and turn on a top-loader balance. Place a clean, dry 150-mL beaker on the balance and zero it.

1-4. With the beaker still on the balance, add to the beaker 2.0 ± 0.1 g of $CuCl_2 \cdot 2H_2O$, copper(II) chloride dihydrate by gently tapping a spatula containing the crystals with your free hand. Record the mass. Leave the balance and the area around it clean.

CAUTION

Copper(II) chloride solution is toxic by ingestion and is a body tissue irritant; avoid contact with your skin. Wash your hands before leaving the laboratory.

1-5. Add 100.0 mL of distilled water to the beaker with a graduated cylinder. Swirl the beaker and stir until the crystals are dissolved. Observe the appearance of the solution.

1-6. Use a clean and dry 25-mL graduated cylinder to measure 17.5 mL of this copper chloride solution into each of the test tubes labeled 1 to 4.

1-7. Clean and dry your 25-mL graduated cylinder.

1-8. Your instructor will provide you with an unknown copper(II) solution for tube 5. You do not know the molarity of this solution. Use your clean and dry 25-mL graduated cylinder to measure 17.5 mL of this copper chloride solution into the test tube labeled 5.

1-9. Obtain five pieces of paper. Label these with the numbers1 to 5.

1-10. Place paper #1 on the balance and zero it. Measure aluminum wire in the amount given in the table and record the exact mass. Then repeat for papers #2 to 5.

Paper/Test Tube	Mass Needed +/− 0.01 g	Approximate Length
1	0.03	1.2 cm
2	0.07	2.6 cm
3	0.14	5.2 cm
4	0.20	7.3 cm
5	0.10	3.8 cm

1-11. Put the aluminum wire into the appropriate test tube. Record observations of each tube.

1-12. Place the five tubes into the boiling water bath. You will need to adjust the heat to keep the water bath at a slow boil. You will need to stir the tubes to remove any deposits after each minute of heating. You can use a stirring rod or a thin spatula to do this. Continue to stir as any deposits build up. Keep the tubes in the bath for approximately 25 to 30 minutes. At least one of the solutions should become completely colorless.

1-13. Turn off the heat and observe for 10 to 15 more minutes, stirring as needed. Then remove the tubes from the boiling water bath and let them cool. Record your observations. You should examine the tubes over white paper to be able to detect colors.

1-14. If there is any remaining wire in a test tube, record the number of the test tube. Carefully remove the wire. Use a thin spatula to trap the wire against the side of the tube. Try to keep any residue in the test tube. Transfer any remaining residue from the wire back to the tube with a spatula.

1-15. Gently dry the wire by blotting it with a paper towel. Measure and record the mass of the dry left over aluminum wire.

1-16. Obtain a piece of filter paper and a watch glass. Put the tube number and your initials on the filter paper. Record the mass of the filter paper and the watch glass.

1-17. Set up a suction filtration. (See the Common Procedures and Concepts section at the end of this manual.) Line the funnel with a filter paper that has been moistened with distilled water. Turn on the aspirator. Decant the supernatant liquid through the funnel. Transfer the residue to the funnel. With a wash bottle of distilled water, rinse the residue three times with about 5 to 10 mL each time.

1-18. With a wash bottle of ethanol, rinse the residue three times with about 5 mL each time. Water will be removed from the residue, as it is soluble in ethanol.

CAUTION

Alcohol and acetone are very flammable. There should be no open flames nearby.

1-19. Finally, with a wash bottle of acetone, rinse the residue three times with about 5 mL each time. Disconnect the hose. Turn the aspirator off.

1-20. Carefully remove the filter paper, taking care not to tear it. Place the filter paper on a watch glass.

1-21. Place the watch glass under a heat lamp for about 10 minutes or until dry. You may also use an oven just over 100°C. Overheating will cause oxidation to a green or dark brown.

1-22. While you are drying the contents of tube 1, repeat Procedures 14 to 21 for the other test tubes.

1-23. Once the filter paper is dry. Remove the watch glass from the oven. Let the watch glass cool. Weigh the watch glass, filter paper, and residue. Do this for all five watch glasses.

1-24. Clean and dry all of the glassware. Make sure that your work area and the balance you used are clean.

6 EXPERIMENT 6: ALUMINUM AND COPPER(II) CHLORIDE

Prelab Exercises

1. Describe the purpose of a hot water bath.

2. The name for $CuCl_2 \cdot 2H_2O$ is copper(II) chloride dihydrate. Why are the roman numerals in the name?

3. How many moles are in 4.2 g of $CuCl_2 \cdot 2H_2O$?

4. What is the molarity of 4.2 g of $CuCl_2 \cdot 2H_2O$ that is mixed with 100. mL of water assuming that you produce 100. mL of solution?

5. How many moles are in 30.0 mL of the solution from #4?

6. Describe the advantages of using a suction filtration apparatus. (See the Common Procedures and Concepts section.)

7. Answer any question added by your instructor.

Date _____   **Student's Signature** _____

Name (Print) ______ Date (of Lab Meeting) ______ Instructor ______

Course/Section ______ Partner's Name (If Applicable) ______

6 EXPERIMENT 6: ALUMINUM AND COPPER(II) CHLORIDE

Report Form

DATA

Mass of $CuCl_2 \cdot 2H_2O$: ______ g

Tube 1:

Mass of Al: ______ g

Observations before the hot water bath:

Observations after the hot water bath:

Mass of any remaining Al: ______ g

Mass of filter paper and watch glass: ______ g

Mass of dried filter paper, watch glass, and residue: ______ g

Mass of residue: ______ g

Tube 2:

Mass of Al: ______ g

Observations before the hot water bath:

Observations after the hot water bath:

Mass of any remaining Al: ______ g

Mass of filter paper and watch glass: ______ g

Mass of dried filter paper, watch glass, and residue: ______ g

Mass of residue: ______ g

Tube 3:

Mass of Al: _____ g

Observations before the hot water bath:

Observations after the hot water bath:

Mass of any remaining Al: _____ g

Mass of filter paper and watch glass: _____ g

Mass of dried filter paper, watch glass, and residue: _____ g

Mass of residue: _____ g

Tube 4:

Mass of Al: _____ g

Observations before the hot water bath:

Observations after the hot water bath:

Mass of any remaining Al: _____ g

Mass of filter paper and watch glass: _____ g

Mass of dried filter paper, watch glass, and residue: _____ g

Mass of residue: _____ g

Tube 5:

Mass of Al: _____ g

Observations before the hot water bath:

Observations after the hot water bath:

Mass of any remaining Al: _____ g

Mass of filter paper and watch glass: _____ g

Mass of dried filter paper, watch glass, and residue: _____ g

Mass of residue: _____ g

Date _____ **Instructor's Signature** _____

ANALYSIS

1. Propose a balanced equation for the combination of aluminum with copper(II) chloride dihydrate. Water will be one of your products.

2. Explain what evidence you have that a chemical reaction did occur. (If you saw any bubbling, it was from a second reaction that you do not have to account for in your proposed equation.)

3. Draw a particle view of the main reaction.

4. For tubes 1 to 4, predict which reactant still existed after the water bath. Use your observations as evidence.

Tube	Left Over Reactant	Evidence
1		
2		
3		
4		

5. Find the molarity of the $CuCl_2 \cdot 2H_2O$ solution you created and used in tubes #1 to 4, assuming that you produced 100.0 mL of solution. Show your work here.

6. You used 17.5 mL of this solution in tubes 1 to 4. Calculate the number of moles. Show your work.

7. Fill in the following table. Show a sample calculation for one tube below.

Tube	Moles of $CuCl_2 \cdot 2H_2O$ Initially	Mass of Al Added	Moles of Al Added	Mass of Cu Produced	Moles of Cu Produced
1					
2					
3					
4					

8. For tubes 1 to 4, was one of the reactants completely used up (the limiting reactant)? First predict the limiting reactant by using your observations. Then find the limiting reactant by assuming that your equation is correct and then calculating which reactant should run out first based on the copper produced. Explain your reasoning.

Tube	Limiting Reactant Observation	Limiting Reactant Calculation
1		
2		
3		
4		

9. Find the number of moles of Cu produced in tube 5. Show your calculations below.

Moles of Al Added	Mass of Cu Produced	Moles of Cu Produced

10. Find the moles of $CuCl_2 \cdot 2H_2O$ that reacted in tube 5. Describe your rationale and show your work below.

11. Find the molarity of the $CuCl_2 \cdot 2H_2O$ solution used in tube 5. Show your work here.

POSTLAB QUESTIONS

1. What is a limiting reactant?

2. When determining the limiting reactant in a reaction is it enough to just use the mass of each reactant without any conversion? Explain.

3. Describe the relationship between how much reacts and the balanced equation.

4. Using the following unbalanced equation and information, find the limiting reactant. Explain your thinking.

$$\_\_\_\_\_NH_3 + \_\_\_\_\_O_2 \rightarrow \_\_\_\_\_NO + \_\_\_\_\_H_2O$$

21.5 g of NH_3 and 55.3 g of oxygen gas

5. For the following reactions, identify the limiting and excess reactants. Give the **reasons** for your choices.

a) The combustion of a candle in an open area.

b) The addition of a teaspoon of baking soda in a bottle of vinegar.

c) Spoons that are made out of sterling silver tarnish when they are in contact with sulfur in the air.

Date _____ **Student's Signature** _____

Recycling Aluminum Cans

A Skill Building Experiment

INTRODUCTION

Aluminum is the third most abundant element in the earth's crust; however, the supply of useful ores rich in aluminum is limited. Manufacturing aluminum from its ores consumes large quantities of electrical energy. Recycle programs for aluminum cans are helping to reduce the amount of electrical energy needed and, therefore, reduce the cost of many aluminum products. In this experiment you will recycle aluminum from a used beverage can by converting it into another useful product.

You will prepare "alum" from the waste aluminum can. The alum you prepare will be potassium aluminum sulfate dodecahydrate, $KAl(SO_4)_2 \cdot 12H_2O$. Alums are ionic compounds containing a sulfate anion, a trivalent cation (e.g., Al^{3+}, Cr^{3+}, Fe^{3+}), and a monovalent cation (e.g., K^+, Na^+, NH_4^+). The potassium aluminum sulfate dodecahydrate is the most commonly encountered alum. Various forms of alum are widely used in the manufacture of pickles and leather, in water purification and wastewater treatment plants as coagulants or flocculants, and in the fabrics industry as binders.

OBJECTIVES

You will perform the synthesis of alum from aluminum. You will need to make stoichiometric calculations and prepare the needed portion of an aluminum can and the solutions necessary to convert the aluminum into alum. You must safely transfer and heat an alkaline solution, safely neutralize the alkaline solution, and precipitate and isolate the product. Yield and percent yield will be calculated based upon the aluminum being the limiting reactant. This experiment should help you to perfect your basic laboratory skills, such as weighing, heating solutions, preparing solutions, isolating products, etc.

CONCEPTS

Aluminum is a "reactive" metal, but it reacts very slowly with dilute acids because its surface is a very thin, impenetrable coating of aluminum oxide. Alkaline solution (containing OH^-), however, attacks the metal after dissolving the oxide layer. When aluminum reacts with KOH, it is oxidized to form the potassium tetrahydroxoaluminate(III), $K[Al(OH)_4]$, which is stable only in the basic solution.

NOTE: You will not be expected to be able to name complexes like this one, unless specifically told otherwise by your instructor.

The formula unit equation for this first reaction in the series of reactions that you will perform in this experiment is:

$$2Al(s) + 6H_2O(l) + 2KOH(aq) \rightarrow 2K[Al(OH)_4](aq) + 3H_2(g)$$

When sulfuric acid is slowly added to the resulting alkaline solution of this initial product, a portion of the acid acts to neutralize the excess potassium hydroxide and to convert it to potassium sulfate and water. The formula-unit equation for this reaction is:

$$2KOH(aq) + H_2SO_4(aq) \rightarrow K_2SO_4(aq) + 2H_2O(l)$$

As additional sulfuric acid is added, one hydroxide ion is removed from each potassium tetrahydroxoaluminate(III) causing the precipitation of white, gelatinous, semi-solid aluminum hydroxide, $Al(OH)_3$.

$$2K[Al(OH)_4](aq) + H_2SO_4(aq) \rightarrow 2Al(OH)_3(s) + K_2SO_4(aq) + 2H_2O(l)$$

When more sulfuric acid is added, the aluminum hydroxide dissolves to form a warm solution of soluble aluminum sulfate.

$$2Al(OH)_3(s) + 3H_2SO_4(aq) \rightarrow Al_2(SO_4)_3(aq) + 6H_2O(l)$$

The warm aluminum sulfate solution also contains potassium sulfate from the previous step. If the solution is cooled, potassium aluminum sulfate dodecahydrate or "Alum" precipitates. Sometimes a seed crystal is needed to initiate the precipitation.

$$Al_2(SO_4)_3(aq) + K_2SO_4(aq) + 24H_2O(l) \rightarrow 2K[Al(SO_4)_2] \cdot 12H_2O(s)$$

If the equations for the above steps are combined, a single overall equation that represents the overall chemical conversion of aluminum to alum is formed.

$$2Al(s) + 2KOH(aq) + 4H_2SO_4(aq) + 22H_2O(l) \rightarrow$$
$$2K[Al(SO_4)_2] \cdot 12H_2O(s) + 3H_2(g)$$

TECHNIQUES

Lighting and adjusting a burner, working with hot and combustible materials, transferring and mixing of reactive solutions, using the aspirator for vacuum filtration, and disposing of reaction solutions and product are just some of the techniques utilized.

ACTIVITIES

You will make alum from an aluminum can and perform stoichiometry calculations needed in order to determine amounts of reactants to use and the yield obtained.

CAUTION

During this experiment, you will be working with strong acids and bases. You must wear approved eye protection and proper protective clothing at all times. Alkaline solutions are particularly hazardous to the eyes. In case any of the reagents used in this experiment come in contact with your skin or eyes, wash the affected area immediately with lots of water. Notify your instructor. Exercise special care while hydrogen, a flammable gas, is being formed. (See the CAUTION statement in Procedure 1-3.)

PROCEDURES

1-1. Being careful not to cut yourself on the sharp edges of the can, use scissors to cut a piece approximately 2 inches × 2 ½ inches from an aluminum can. Remove most of the paint and inside coating from the piece you just cut out. You may need to use the scissors to scrape off the paint.

NOTE: A considerable amount of time will be saved if you scrape off the paint and inside coating from a portion of a beverage can before you bring the can to the laboratory.

1-2. Cut your aluminum sample into small pieces about ¼ inch on each side. Place the pieces in a 250-mL beaker. On a top-loader balance tare a clean sheet of paper. Transfer between 1.00 and 1.20 g of your cut pieces onto the paper and weigh. Record in your laboratory notebook the mass of the pieces selected. Discard the excess pieces and the remainder of the can in a container designed for "sharps." Transfer the weighed pieces of aluminum back into the beaker.

1-3. *If possible, work at a fume hood.* To the pieces of aluminum in the 250-mL beaker add a volume of 1.5 *M* potassium hydroxide that is 1.5 times the volume that you calculated in question #4 on the Prelab.

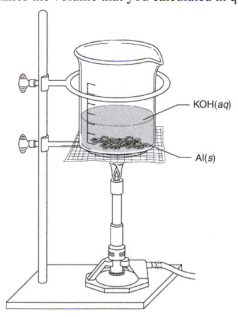

KOH(*aq*)

Al(*s*)

CAUTION

Bubbles of hydrogen should soon form as the result of the reaction between aluminum and aqueous hydroxide. Heat the beaker very gently to speed up the reaction, preferably using a hot plate. If a burner is used, be very careful. Hydrogen-air mixtures can ignite explosively. However, hydrogen is also much lighter than air. If allowed to escape, hydrogen will leave the area very quickly and form a mixture with air that is too dilute to ignite.

When the liquid level in the beaker drops to less than half of its original volume, add distilled water to maintain the volume at slightly less than half of its original volume. The reaction should take no more than 30 minutes. When the hydrogen evolution ceases, the reaction is complete.

NOTE: In the course of the reactions, the mixture will turn from colorless to dark gray or black. Decomposition of residual paint or plastic lining is most likely the source of the dark material. Periodically, aluminum fragments will rise and fall during the reaction. Pieces of plastic lining may remain after the reaction has stopped.

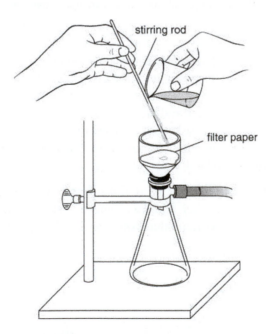

stirring rod

filter paper

1-4. Remove any solid residue from the reaction solution. Use an aspirator and vacuum filtration apparatus to filter the hot solution. Consult the Common Procedures and Concepts Section at the end of this manual if you are unfamiliar with the technique of vacuum filtration. Securely clamp the vacuum flask and moisten the filter paper before you begin. Filter the solution. After all of the liquid has passed through the filter paper, disconnect the rubber tubing from the filter flask to break the vacuum.

CAUTION	

Do not turn off the aspirator until the vacuum has been broken.

1-5. Any dark residue should be left on the filter paper and the filtrate should be **clear**. With the aspirator on, rinse the beaker twice with 5-mL portions of distilled water, pouring each rinse through the filter residue. Break the vacuum, and then turn off the aspirator.

Your solution may be slightly **colored** at this point. *Clear and colorless are not the same things!*

1-6. Get 20 mL of 9.0 M H_2SO_4. If it is not provided, prepare it by placing crushed ice and then distilled water into a container to bring the level of the ice/water mixture to the 10-mL mark. *Use only a container made of* Pyrex. **Very carefully** pour small portions of concentrated (18 M) sulfuric acid from a small reagent bottle into the water and ice. Use a glass rod to stir after each small addition. Continue until the liquid level reaches the 20-mL mark on the container. Stir the solution thoroughly, but carefully. *The diluted acid will be quite hot! When diluting concentrated acids, one should always stir while adding the concentrated acid to the water. Considerable heat will be generated.*

CAUTION	

9.0 M H_2SO_4 is still a very corrosive acid that can cause severe damage to clothing, skin, etc.

1-7. Pour the clear filtrate into a clean 250-mL beaker. To transfer all the product, rinse the filter flask with 10 mL of distilled water and pour the rinse water into the beaker. Place the beaker in a cooling bath if the filtrate is not yet cool. *Slowly and carefully, while stirring,* add 20 mL of 9.0 M H_2SO_4 to the cooled solution. *A white precipitate of aluminum hydroxide should soon appear. The neutralization reaction will also generate notice -able heat. Adding the last few milliliters of the sulfuric acid will dissolve dissolve the $Al(OH)_3$.*

If necessary, warm the solution gently, while stirring, to completely dissolve any undissolved $Al(OH)_3$. The final solution will contain potassium ions (from the KOH used), aluminum ions, and sulfate ions. If any solid residue remains after a few minutes of heating, filter the mixture and work with the clear filtrate.

1-8. Fill a 1-L plastic beaker half way with crushed ice to make an ice bath. Add water to just cover the ice. Place the reaction beaker (from Procedure 1-7) in the ice-water bath to chill. Crystals of the alum should begin to form. Chill the mixture thoroughly for about 15 minutes. If crystals do not form, (a) reduce the volume of solution by boiling away some of the water or (b) induce crystallization (consult your instructor). To induce crystallization, you can add one or two very tiny seed crystals. Seed crystals can be made by placing a drop of solution on the end of a stirring rod and blowing on it until it is dry.

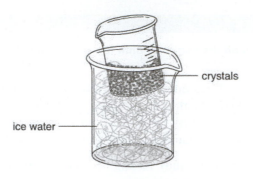

1-9. Clean and reassemble the vacuum filtration equipment. Filter the alum crystals from your chilled solution, transferring as much of the crystalline product as possible to the funnel.

1-10. Mix and chill 12 mL of 95% ethanol with 12 mL water for a few minutes in the ice-water bath. Rinse the remaining crystals from the beaker into the funnel with half of this solution. Repeat with the second half of the solution. Spread the crystals evenly on the filter paper with your spatula. Allow the aspirator to pull air through the crystals for about 10 minutes. *Ethanol in the rinse solution reduces the solubility of the alum.*

1-11. Tare a sheet of weighing paper on a top-loading balance. Use your spatula to transfer all of the air-dried crystals from the filter paper onto the tared weighing paper. Weigh the product.

1-12. Show your alum to your instructor, report the mass of alum obtained, and request a "product inspection." If your alum passes inspection, transfer the alum into the "Alum Storage Bottle" for later use. Otherwise, dispose of the alum as directed by your instructor.

7   EXPERIMENT 7: RECYCLING ALUMINUM CANS

Prelab Exercises

1. Give the chemical formula of the alum produced in this investigation. Give its formula weight or molar mass.

2. Give the chemical formula of an alum other than the one produced in this investigation.

3. Write the formula unit equation for the formation of "alum" from aluminum.

4. According to the equation, $2Al(s) + 6H_2O(l) + 2KOH(aq) \rightarrow 2K[Al(OH)_4](aq) + 3H_2(g)$, what minimum volume of 1.5 M KOH would be needed to react with 1.10 g of Al? (Record this question and answer in your notebook.)

5. According to the equation, $2Al(s) + 6H_2O(l) + 2KOH(aq) \rightarrow 2K[Al(OH)_4](aq) + 3H_2(g)$, how many grams of hydrogen gas would be formed in the reaction of 1.10 g of Al and excess KOH?

6. Using your own words, define molarity (M) in terms of grams and formula weights (or molar masses).

7. Draw particle views of the reactants and products in the single overall equation for the conversion of aluminum to alum. Label each as a dissolved salt, strong acid, gas, solid, etc.

Date _____   **Student's Signature** _____

Name (Print)   Date (of Lab Meeting)   Instructor

Course/Section   Partner's Name (If Applicable)

7 EXPERIMENT 7: RECYCLING ALUMINUM CANS

Report Form

DATA

Mass of aluminum used: _____ g

Volume of 1.5 M KOH used: _____ mL

Volume of 9.0 M H_2SO_4 used: _____ mL

Mass of product obtained: _____ g

Instructor's evaluation of product   Technical grade ☐   Reagent grade ☐

Date _____ **Instructor's Signature** _____

ANALYSIS

Moles of aluminum used: _____ moles   Moles of H_2SO_4 used: _____ moles

Moles of KOH used: _____ moles

Theoretical yield of alum based upon aluminum being the limiting reactant in the overall reaction equation: _____ moles = _____ g

Yield: _____ %

POSTLAB QUESTIONS

1. If your product had not been dry, would the yield be higher or lower than the yield based on a dry product? Explain your choice.

2. How many grams of alum can be obtained from 22.0 g of aluminum when the reaction proceeds with 100% yield? How many grams of alum would be obtained if the reaction were to proceed with 75.0% yield?

3. Why was excess KOH used?

4. Explain why boiling the alum solution would help induce crystal formation.

5. Describe three techniques that were used in this experiment.

6. How would you modify the current procedures to produce the product: $NaCr(SO_4)_2 \cdot 12H_2O(s)$?

Date _____ **Student's Signature** _____

Patterns

A Guided Inquiry Experiment

INTRODUCTION

The periodic table of elements contains all the known elements, both natural and synthetic. A column in the periodic table is called a family or a group, while a row is called a period. There have been a number of proposals for different arrangements of the periodic table.

OBJECTIVES

In this experiment, you will investigate the reactivity of elements with water and dilute acid, then look for any patterns that exist, including any relationships with the periodic table. You will use relationships to create your own periodic table.

CONCEPTS

This experiment uses the concepts of evidences of a chemical change and balancing equations. Blue litmus paper turns pink when hydrogen ions are present, and red litmus paper turns blue when hydroxide ions are present. Phenolphthalein turns pink when hydroxide ions are present.

TECHNIQUES

Observation, collecting a gas above water, and testing for the identity of a gas are used. You may wish to review the Qualitative Tests of Common Gases in the Common Procedures and Concepts Section at the end of this manual.

ACTIVITIES

You will investigate the possible reactions of a number of elements with water and with a dilute solution of hydrochloric acid. Next you will propose trends in the reactivity based on the element's location in the periodic table. Finally, you will devise and explain your own periodic table for elements from a fictional planet.

PROCEDURES

1-1. After observing the demonstrations by your instructor, record the data from the demonstration of sodium, lithium, and potassium.

1-2. Obtain a clean 400-mL beaker and an 18 × 150-mm test tube. Half fill the beaker with distilled water. Fill the test tube with distilled water. Holding your fingers over the mouth of the test tube, invert it into the beaker. Do not allow any air bubbles into the test tube.

1-3. Obtain a piece of calcium, Ca. Record observations of the appearance of calcium. Place the Ca into the beaker. Collect any gas given off until you have filled the test tube approximately half way.

1-4. Discuss with your partner what gas(es) are possible with this system at room temperature (you may have done this during your prelaboratory discussion). Test the gas using the procedures outlined in the Common Procedures and Concepts Section at the end of this manual. If necessary, collect more gas by repeating Procedures 1-2 and 1-3. Record your observations.

1-5. Test the liquid in the beaker with litmus paper.

1-6. Obtain three clean 18 × 150-mm test tubes.

1-7. Obtain pieces of magnesium, Mg, and aluminum, Al. Also obtain enough sulfur to cover the end of a spatula tip. Make observations of the appearance of these elements.

1-8. Place these elements in separate test tubes. Fill the tubes half way with distilled water and investigate the interactions. Record your initial observations. Then put the test tubes into a test tube rack or beaker to observe again later.

1-9. Obtain three additional clean 18 × 150-mm test tubes.

1-10. Obtain pieces of magnesium, Mg, and aluminum, Al. Also obtain enough sulfur to cover the end of a spatula tip. Place these elements in separate test tubes. Fill the tubes half way with 0.10 *M* HCl and investigate the interactions. Record your initial observations. Then put the test tubes into a test tube rack or beaker to observe again later.

1-11. Record final observations of the test tubes containing water and those containing HCl.

1-12. Empty the solutions in the waste beaker and clean all glassware.

8   **EXPERIMENT 8: PATTERNS**

Prelab Exercises

1. What is the difference between a group and a period in the periodic table?

2. What is an element?

3. List three elements you use and give their uses.

4. Why is it necessary to have clean glassware before beginning an experiment?

5. List three observations that may lead you to conclude there has been a chemical reaction.

Date _____   **Student's Signature** _____

Name (Print) _____ Date (of Lab Meeting) _____ Instructor _____

Course/Section _____ Partner's Name (If Applicable) _____

8   EXPERIMENT 8: PATTERNS

Report Form

DATA

Observations in Distilled Water

Element	Observations of the Element	Observations of the Combination with Water
Na		
Li		
K		
Ca		
Mg		
Al		
S		

Observations in Dilute Acid

Element	Observations of the Combination with Acid
Mg	
Al	
S	

Date _____ **Instructor's Signature** _____

ANALYSIS

1. Did a chemical reaction or a physical mixture take place when you combined the calcium with water? Explain your reasoning.

2. Write a balanced chemical equation for the reaction(s) with water that you observed. Include any reactions for calcium, sodium, potassium, lithium, magnesium, aluminum, or sulfur.

3. Did a chemical reaction or a physical mixture take place when you combined the elements with a dilute acid? Explain your reasoning.

4. Assuming that the identity of any gas produced was the same as with water, write a balanced chemical equation for the reaction(s) you observed with acid.

5. Describe any patterns you see in reactivity as you go across a row (a period) in the table.

6. Locate the elements that you investigated on the periodic table. Identify any patterns in reactivity as you go down a column (a family or group) in the table.

Explore the patterns in the elements as you go across the rows and down the columns in the periodic table for the atomic and ionic radii in the tables you are given. Discuss this with your partner. You will be asked to give the patterns you found, and the rationale you used to propose the pattern(s). Use the following tables to answer questions 7 and 8.

Alkali Metals

Element	Symbol	Atom Radius pm	+1 Ion Radius pm
Lithium	Li	152	90
Sodium	Na	186	116
Potassium	K	227	152
Rubidium	Rb	248	166
Cesium	Cs	265	181

Alkaline Earth Metals

Element	Symbol	Atom Radius pm	+2 Ion Radius pm
Beryllium	Be	112	59
Magnesium	Mg	160	85
Calcium	Ca	197	114
Strontium	Sr	215	132
Barium	Ba	222	149

Oxygen Group

Element	Symbol	Atom Radius pm	−2 Ion Radius pm
Oxygen	O	73	126
Sulphur	S	103	170
Selenium	Se	119	184
Tellurium	Te	142	207
Polonium	Po	168	

Halogens

Element	Symbol	Atom Radius pm	−1 Ion Radius pm
Fluorine	F	72	119
Chlorine	Cl	100	167
Bromine	Br	114	182
Iodine	I	133	206
Astatine	As	140	

Noble Gases

Element	Symbol	Atom Radius pm	
Helium	He	31	
Neon	Ne	71	
Argon	Ar	98	
Krypton	Kr	112	
Xenon	Xe	131	
Radon	Rn	141	

Third Row Elements

Element	Symbol	Atom Radius pm	Ion Radius pm
Sodium	Na	186	(+1) 116
Magnesium	Mg	160	(+2) 85
Aluminium	Al	143	(+3) 68
Silicon	Si	118	
Phosphorous	P	110	
Sulphur	S	103	(−2) 170
Chlorine	Cl	100	(−1) 167
Argon	Ar	98	

7. Identify any patterns in atomic radii as you go down a column (a family or group) and as you go across a row (a period) in the table. Give your rationale.

8. Identify any patterns in ionic radii as you go down a column (a family or group) and as you go across a row (a period) in the table. Give your rationale.

Explore the patterns in the elements as you go across the rows and down the columns in the periodic table for the various ionization energies in the tables you are given. Discuss this with your partner. You will be asked to give the patterns you found and the rationale you used to propose the pattern(s). Use the following tables to answer questions 9 and 10.

Alkali Metals (Ionization Energy in kJ/mol)		IE +1	IE +2	IE +3	IE +4	Alkaline Earth Metals (Ionization Energy in kJ/mol)		IE +1	IE +2	IE +3	IE +4
Element	Symbol					Element	Symbol				
Lithium	Li	520	7298	118151	-	Beryllium	Be	899	1757	14848	21006
Sodium	Na	496	4562	6912	9543	Magnesium	Mg	738	1451	7732	10540
Potassium	K	419	3051	4411	5877	Calcium	Ca	599	1145	4910	6491
Rubidium	Rb	403	2632	3859	5080	Strontium	Sr	550	1064	4138	5500
Cesium	Cs	377	2234	3400		Barium	Ba	503	965	3600	

Third Row Elements (Ionization Energy in kJ/mol)

Element	Symbol	IE +1	IE +2	IE +3	IE +4	IE +5	IE +6	IE +7	IE +8	IE +9
Sodium	Na	496	4562	6912	9543	13354	16613	20117	25496	28932
Magnesium	Mg	738	1451	7732	10540	13630	18020	21711	25661	31653
Aluminium	Al	578	1817	2745	11577	14842	18379	23326	27465	31853
Silicon	Si	786	1577	3231	4354	16091	19805	23780	29287	33878
Phosphorous	P	1012	1903	2912	4950	6274	21267	25431	29872	35905
Sulphur	S	1000	2251	3361	4565	7004	8496	27107	31719	36621
Chlorine	Cl	1251	2297	3826	5160	6542	9362	11018	33604	38600
Argon	Ar	1521	2665	3928	5770	7238	8781	11995	13842	40760

9. Identify any patterns in the first ionization energy as you go down a column (a family or group) and as you go across a row (a period) in the table. Give your rationale.

10. What additional pattern(s) can you find in the ionization energy for the third period? Consider all patterns, including where you have the largest increase. Give your rationale.

POSTLAB QUESTIONS

1. Is there a correlation between reactivity of elements and their position in the periodic table? Explain.

2. Predict what would occur if the following were placed in water. Compare their reactivity as greater or less than the elements that you observed.

Beryllium (Be)

Strontium (Sr)

Silicon (Si)

Cesium (Cs)

3. What would you predict about the reactivity of potassium in hydrochloric acid compared to potassium in water?

4. Which atom would you expect to have the largest radius, carbon or nitrogen? Explain your rationale.

5. Which atom would you expect to have the largest radius, silicon or germanium? Explain your rationale.

6. Which atom would you expect to have the largest first ionization energy, silicon or germanium? Explain your rationale.

7. Based on your findings, between which ionization energies would you expect to have the largest increase for carbon? Consider the 1st, 2nd, 3rd, 4th, 5th, 6th, or 7th ionization energies. Explain your rationale.

8. Based on your findings, between which ionization energies would you expect to have the largest increase for gallium? Consider the 1st, 2nd, 3rd, 4th, 5th, 6th, or 7th ionization energies. Explain your rationale.

9. Assume that you are on a new planet, Planet Kyle. You are delighted to discover that Planet Kyle also contains water, so the elements hydrogen (H) and oxygen (O) exist. However, all other elements are new.

   Based on the following information and assuming the laws of nature are the same as those on Earth, construct a periodic table for Planet Kyle. Remember that you can **only consider relationships in columns or in rows of your periodic table**. All relationships below represent those in columns or in rows. For example #3 is in the same column or row as #2 and #8. This means that #3 is above #2 if they are in the same column or #3 is to the right of #2 if in the same row.

   Fill in the table below AND attach an explanation for your reasoning concerning how you placed the elements. It will help to cut out the element squares below so that you and your partner(s) can easily make changes.

   Experimental Data:

Element 1	Has the lowest 1st IE and largest at. radius in the table
Element 2	Has the largest jump in IE between the 1st and 2nd IE
Element 3	Smaller at. radius than 2 or 8
Element 4	Reacts more than 2 but less than 8, 10, or 1; has the largest IE jump between the 1st and 2nd IE
Element 5	Reacts more than 12 but less than 7
Element 6	Has a smaller at. radius than 1 and more than 7
Element 7	Has a larger 1st IE than 6 and less than 10
Element 8	Less reactive than 7; smaller 1st IE than 3
Element 9	Has smaller at. radius than 12; smaller 1st IE than oxygen
Element 10	Less reactive than 1; larger at. radius than 7
Element 11	Less reactive than 12; larger at. radius than oxygen
Element 12	Has a larger 1st IE than 5; largest IE jump between the 3rd and 4th IE

   (IE = ionization energy; at. radius = atomic radius)

Element 1	Element 2	Element 3	Element 4
Element 5	Element 6	Element 7	Element 8
Element 9	Element 10	Element 11	Element 12

Periodic Table for Planet Kyle

H			
			O
			▨
		▨	▨

10. Which element on Kyle would you expect to be the most reactive in hydrochloric acid? Explain.

Date _____ **Student's Signature** _____

Solutions and Crystals of Alum

A Guided Inquiry Experiment

INTRODUCTION

Solutions are often made from crystalline solids and water. Dissolving of a solid is considered a physical change because the solute can be recovered through a number of physical means. Solutions can have variable concentrations.

Many crystals possess esthetic, or functionally useful optical, electrical, and magnetic properties. Even if the desired crystals are found in nature, nature frequently does not furnish either the quantity, nor the quality of crystals desired. Therefore, many crystals are produced commercially. The relationships between composition, structure, and properties are concerns of many chemists and physicists. The formation of crystals is also used to purify substances in the process known as recrystallization. Recrystallization of a solid involves dissolving the solid in a suitable solvent at an elevated temperature. After any undissolved material is removed by filtration, the filtrate is slowly cooled to allow crystals to form. The substance will be purer if the crystals form very slowly.

This experiment investigates the formation of saturated solutions and demonstrates crystal growth from saturated aqueous solutions.

OBJECTIVES

You will investigate the solubility of sodium thiosulfate in various situations and prepare a saturated alum solution to grow a large single crystal as the solvent slowly evaporates. The growing of the crystal will take place over several days. The result will be a very large, very pure alum or chrome alum or mixed crystal.

CONCEPTS

Many substances will dissolve in water to form aqueous solutions. A solution that contains the maximum amount of a substance that can be made to dissolve without heating the solution is called a saturated solution. The solubility of the substance is known as mass of substance present in a saturated solution prepared with 100 g of water. Solubility can also be expressed in other units.

This process of nucleation may occur when the ions of the solute collide with appropriate orientation and with sufficiently low kinetic energy to permit them

103

to "stick" to each other and to begin crystallization. Frequently, "foreign" solids (irregularities on the wall of the container or dust particles) will serve as nuclei for the formation of crystals. Once a tiny crystal has formed, it will grow when ions hit the faces of the crystal and join the orderly array of ions. To keep the crystal growing, the solution must be cooled to even lower temperatures or more solvent must be evaporated. To grow large crystals, small seed crystals are first prepared. A well-formed seed crystal is suspended in a saturated growing solution and the solvent is slowly evaporated. A huge, perfectly octahedral crystal of alum can be obtained by replenishing the growing solution.

TECHNIQUES

Preparing a hot water bath, preparing solutions, and growing crystals are just some of the techniques utilized.

ACTIVITIES

You will prepare a saturated solution and observe it during changes in temperature. Then you will produce a very large and attractive crystal of alum.

CAUTION

You must wear approved eye protection at all times. Alum solutions are strong astringents. Dispose of unwanted sodium thiosulfate and alum solutions according to directions provided by your instructor. Rinse all containers with copious amounts of water. Store your growing solutions in a place in which they will not be disturbed.

PROCEDURES

Solutions

1-1. Record the temperature in the laboratory. Select, turn on, and zero a top-loader balance. Place a clean, dry 100-mL beaker on the balance. Record the mass of the beaker. Place 15 ± 0.20 g of sodium thiosulfate, $Na_2S_2O \cdot 5H_2O$ in the beaker. Record the mass of the beaker plus sodium thiosulfate.

1-2. Transfer about 2 g of the crystals of sodium thiosulfate from the beaker to a test tube. Obtain 3.0 mL of distilled water in a 10-mL graduated cylinder. Transfer the water to the test tube with the sodium thiosulfate.

1-3. Mix the contents of the test tube by holding the test tube near its top with the thumb and index finger of one hand while tapping the side of the test tube near its bottom with the index finger of the other hand. You should observe a swirling of the contents of the test tube.

If the crystals dissolve, add a few more. Continue to slowly add crystals a few at a time until no more will dissolve. If more than a few crystals remain at the bottom of the test tube when no more will dissolve, START OVER. Note any change in temperature as dissolution occurs by touching the bottom of the test tube to your inner wrist. Record your observations. Note: Save this test tube and its contents for Procedure 1-5.

1-4. Take the beaker with the unused sodium thiosulfate in it to the balance. Record the mass of the beaker and unused sodium thiosulfate. Determine the mass of the sodium thiosulfate that has been dissolved in the test tube. The unused sodium thiosulfate will be used in Procedure 1-5.

1-5. Fill a 250-mL beaker ¾ full of water. Place the beaker inside an iron ring and on a wire gauze and iron ring on a ring stand. Stand the test tube containing the solution from Procedure 1-3 in the beaker.

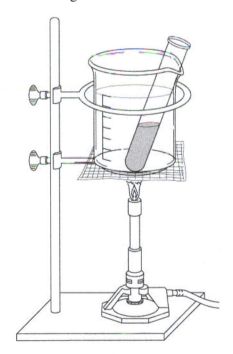

Heat the beaker and its contents with your Bunsen burner. Do NOT allow the solution in the test tube to come to a boil. Add to the solution in the test tube a portion of the sodium thiosulfate remaining in the beaker that was weighed in Procedure 1-4. Continue to warm the test tube and its contents in the water bath. Add large portions of your sodium thiosulfate sample until all the sodium thiosulfate sample is in solution. Stir with a stirring rod to dislodge any of the air bubbles that adhere to the side of the test tube.

1-6. Fill a second 250-mL beaker ¾ full of room temperature water.

Remove the test tube from the hot water beaker and remove the stirring rod. Place the test tube in the beaker of room temperature water. Let the test tube and solution stand undisturbed for about 15 minutes.

1-7. Pick up the test tube and carefully observe its contents. The solution should be clear with no solid visible. Note the temperature qualitatively by touching the bottom of the test tube to your inner wrist.

1-8. If the solution in the test tube is near room temperature and no crystals have formed, drop a tiny crystal of solid sodium thiosulfate into the solution in the test tube and make observations. Make visual observations and qualitative temperature observations every 15 seconds until all action ceases.

1-9. Dispose of the contents of the test tube as your instructor directs.

NOTE: You may need to rewarm the contents of the test tube before pouring it into a collection container.

Crystals of Alum

2-1. Place on a top-loader balance a clean 250-mL beaker. Tare the top-loader balance with the beaker on it. Place approximately 10 g of $KAl(SO_4)_2 \cdot 12H_2O$ [alum] in the beaker. Record the mass of the alum in the beaker. At your work area, add 70 mL of distilled water to the beaker. Heat the mixture and stir until all the alum has dissolved. If the mixture remains cloudy, let it stand for a few minutes to allow the suspended matter to settle out. Then carefully decant the clear solution into another clean 250-mL beaker.

2-2. Tie a piece of thread to a glass rod. Adjust the length of the thread so it does not touch the bottom of the beaker. Place oil on the part of the thread above the solution to prevent the solution from creeping up the thread. Position the rod and thread on top of the beaker such that the thread is in the solution, cover the beaker with a sheet of paper held in place with a rubber band, and store the beaker in a safe place. Within a day or two, crystals should form on the submerged thread, at the bottom of the beaker or in both places. If not, warm the solution to evaporate more solvent and repeat this procedure.

2-3. Take the thread from the solution and remove all the crystals except the best formed one from the thread. If no suitable crystal has formed on the thread, decant the solution from the crystals at the bottom of the beaker. Choose a crystal that has a regular symmetrical shape and smooth faces. Tie a fine thread to the selected crystal. It is a good idea to keep several crystals in reserve, in case something destroys your first-choice seed crystal.

2-4. It may become necessary to prepare a fresh solution of 10 g alum in 70 mL water as described in Procedure 2-1 to add to the original solution. If additional solution is added, allow the new solution to cool to room temperature. Suspend your chosen seed crystal in the solution, cover the beaker with a sheet of paper held in place with a rubber band, and keep the beaker undisturbed.

NOTE: If you add your seed crystal to an unsaturated solution, the crystal will begin to dissolve and may be lost. To prevent such a situation, watch the solution in the vicinity of the seed crystal after you place it in the solution. If the solution is unsaturated and the crystal begins to dissolve, the part of the solution in contact with the crystal will become more concentrated and denser than the solution farther away from the crystal. Denser solution flows toward the bottom of a beaker. If you see such a density current, remove the seed crystal and cool the solution further, or dissolve more alum.

2-5. Over a period of several weeks, remove the crystal and suspend it in a fresh, saturated alum solution. In order to keep large crystals completely submerged, you might have to prepare larger volumes of alum solutions while always maintaining the ratio of 1 g of alum to 7 mL of water. Weigh your crystal and measure its faces once or twice a week. Describe its color, size, and shape. Your crystal will be graded by your instructor when you terminate this experiment. You can keep the crystal after your work has been graded. To preserve it and prevent it from crumbling to a white powder through loss of water, coat it with a clear plastic spray that your instructor will provide.

Name (Print)   Date (of Lab Meeting)   Instructor

Course/Section

Prelab Exercises

1. Explain the difference between solution and saturated solution.

2. Define solvent and solute.

3. Calculate the formula mass of $Na_2S_2O_3 \cdot 5H_2O$.

4. Calculate the formula mass of $KAl(SO_4)_2 \cdot 12H_2O$.

5. Why is the beaker in which the crystal grows covered but not with an air-tight cover?

6. Why is oil or grease applied to a portion of the thread above the solution when growing a crystal?

Date _____ **Student's Signature** _____

9   **EXPERIMENT 9: SOLUTIONS AND CRYSTALS OF ALUM**

Report Form

DATA

Solutions

Temperature in lab: _____

Mass of beaker: _____ g

Mass of beaker + $Na_2S_2O_3 \cdot 5H_2O$ originally obtained: _____ g

Mass of $Na_2S_2O_3 \cdot 5H_2O$ originally obtained: _____ g

Mass of beaker + $Na_2S_2O_3 \cdot 5H_2O$ after dissolving: _____ g

Mass of $Na_2S_2O_3 \cdot 5H_2O$ dissolved in test tube at room temperature: _____ g

Solubility of $Na_2S_2O_3 \cdot 5H_2O$ at room temperature: _____ g/100 mL

Solubility found by others in your laboratory:

_____ g/100 mL   _____ g/100 mL   _____ g/100 mL   _____ g/100 mL

_____ g/100 mL   _____ g/100 mL   _____ g/100 mL   _____ g/100 mL

Observations of the solution:

Observation of any temperature change at dissolution:

111

Observations of the solution after heating, then cooling to room temperature:

Observation of the solution and the temperature every 15 seconds after the addition of the tiny crystal of solid $Na_2S_2O_3 \cdot 5H_2O$:

Crystals of Alum

Record of Crystal Growth

Date	Solution		Mass of Crystal
	g Alum	mL Water	

Submit your crystal for evaluation and grading.

Date _____ **Instructor's Signature** _____

ANALYSIS

Solutions

1. Based upon your data, what is the solubility for $Na_2S_2O_3 \cdot 5H_2O$?

2. Based upon your data and the data obtained from others in your class, what is the solubility for $Na_2S_2O_3 \cdot 5H_2O$?

3. Was the room temperature solution saturated BEFORE you placed the test tube in the hot water and added the remaining $Na_2S_2O_3 \cdot 5H_2O$? Explain why or why not. If it is not saturated, what must happen to the amount of dissolved solute for it to become saturated?

4. How did the dissolving process at room temperature affect the temperature of the water in the solution? How can you account for any change in heat energy?

5. Was the solution saturated AFTER you placed the test tube in the hot water and added the remaining $Na_2S_2O_3 \cdot 5H_2O$? Explain why or why not. If it is not saturated, what must happen to the amount of dissolved solute for it to become saturated?

6. How did temperature affect the solubility?

7. What was the effect of the room temperature water bath on the solution?

8. Was the solution saturated AFTER you cooled the test tube in the room temperature water? Explain why or why not. If it is not saturated, what must happen to the amount of dissolved solute for it to become saturated?

9. How did the temperature vary for the process of crystallization? How can you account for any change in heat energy?

10. Draw a particle view of the crystallizing of sodium thiosulfate.

Crystals of Alum

Describe the various stages of the crystal growth.

Sketch your final crystal.

POSTLAB QUESTIONS

1. What is meant by supersaturated? Did you have a supersaturated solution in this experiment? If so, explain when.

2. Which solution (saturated, unsaturated, or supersaturated) represents a non-stable state? Give reasons for your answers.

3. Water evaporates from the solution as the crystal of alum grows. Does the solution in the vicinity of the crystal remain saturated, unsaturated, or supersaturated? Explain.

4. What does it mean to be supercooled?

5. Why might a dry atmosphere cause your crystal to fracture and to eventually become a white powder?

6.   Describe how you would attempt to grow a crystal of common table sugar.

7.   Calculate the number of moles of alum that were in your final crystal. (Assume that your crystal contained only $KAl(SO_4)_2 \cdot 12H_2O$.)

8.   Crystals have a specific regular shape. What does their shape tell us about how the ions/molecules are arranged at the particle level in the crystal?

9.   Answer all questions assigned by your instructor.

Date _____   **Student's Signature** _____

Analysis of a Carbonated Beverage

A Guided Inquiry Experiment

INTRODUCTION

Many acidic compounds occur in nature. Some examples of naturally occurring acids are the fatty acids, amino acids, and citric acid in foods and the hydrochloric acid of the stomach. Nitrogen and sulfur containing oxyacids are found in acid rain, battery acid, and various cleaning agents. In this experiment, you will determine the citric acid content of a carbonated beverage.

OBJECTIVES

You are to investigate the relationships in the reaction of sodium hydroxide and citric acid. You are to perfect several basic laboratory skills frequently used in the laboratory process known as titration. You are to titrate standard, known, and unknown solutions. An indicator will be used to indicate when the endpoint of the titration is reached. Ultimately, you will use your titration data to calculate the concentration of citric acid in a carbonated beverage.

CONCEPTS

Monoprotic acids have only one acidic hydrogen atom per molecule. Diprotic acids have two and triprotic acids have three acidic hydrogen atoms per molecule. Hydrogens in a formula may or may not be acidic. Only those hydrogens that can be released as hydrogen ions in aqueous solutions are considered to be acidic. The neutralization reaction occurring between a monoprotic acid and sodium hydroxide is described by the following equation.

$$\underset{\text{monoprotic acid}}{HA} \quad + \quad \underset{\text{base}}{NaOH} \quad \rightarrow \quad \underset{\text{salt}}{NaA} \quad + \quad \underset{\text{water}}{H_2O}$$

The neutralization reaction occurring between a diprotic acid and sodium hydroxide is:

$$\underset{\text{diprotic acid}}{H_2A} \quad + \quad \underset{\text{base}}{2NaOH} \quad \rightarrow \quad \underset{\text{salt}}{Na_2A} \quad + \quad \underset{\text{water}}{2H_2O}$$

The reaction with a triprotic acid is similar but requires 3 moles of base and produces 3 moles of water instead of 2. When an alkaline solution is added to a solution of an acid, the OH^- from the added alkaline material and the H^+ ions from the acid combine to form undissociated water. After sufficient base has been added to have all the acidic protons neutralized (the endpoint of the

117

titration), any additional base will rapidly increase the hydroxide ion concentration because H^+ ions are no longer present to combine with OH^-. Phenolphthalein reacts with the excess hydroxide ions and forms a bright red solution. Phenolphthalein, before its reaction with OH^-, is colorless. The appearance of a very pale pink color indicates that the neutralization of the acid is complete and that very small excess of base has converted some of the phenolphthalein into its alkaline or red form. This stage in a titration is referred to as the endpoint.

A measured volume of acidic solution of a known concentration and type can be titrated by using a buret to carefully add a solution of sodium hydroxide to it. The volume of sodium hydroxide solution needed to just reach the endpoint can be measured. The moles of sodium hydroxide consumed in the reaction can be calculated. For monoprotic acids, each mole of sodium hydroxide consumed in the neutralization reaction neutralizes one mole of the monoprotic acid. That is: at the endpoint the number of moles of base equals the number of moles of acid. Since molarity can be written as moles per liter, the following equation can be derived.

$$V_{NaOH} \text{ (in liters)} \times M_{NaOH} = \text{moles}_{NaOH} = \text{moles}_{HA} = V_{HA} \text{ (in liters)} \times M_{HA}$$

During the standardization of your sodium hydroxide solution, the moles of monoprotic acid used and volume of sodium hydroxide solution needed to reach the endpoint will be known. The above equation can be used to calculate the concentration of the sodium hydroxide solution.

In this experiment, you will standardize a sodium hydroxide solution that you will then use to titrate the citric acid in a carbonated beverage. Sodium hydroxide is a rather reactive substance. Sodium hydroxide is one of many compounds that is hygroscopic (they absorb water from the air). The moist surface of the sodium hydroxide pellets will also react with carbon dioxide present in the air. For these reasons, sodium hydroxide is seldom considered to be pure. In this experiment, you will prepare a sodium hydroxide solution by diluting a solution of doubtful concentration. The exact molarity of this solution will be determined by standardization with an acid that can easily be obtained in high purity, will not absorb water, and will not react with carbon dioxide or other constituents of the atmosphere. These are all desirable traits of a primary standard. Potassium hydrogen phthalate (KHP), the mono-potassium salt of the diprotic phthalic acid, is such a primary standard. The amount of KHP required for the standardization of a sodium hydroxide solution is carefully weighed on an analytical balance. The KHP sample is dissolved in hot water, phenolphthalein is added, and sodium hydroxide solution is added until the endpoint is reached. Since KHP is a monoprotic acid, at this point equal number of moles of acid and base have reacted.

potassium hydrogen phthalate (KHP)   + NaOH ⟶   potassium sodium phthalate (KNaP)   + H_2O

The representation above of KHP has a six-sided ring of single and double bonded carbon atoms representing the benzene-like portion of the compound. In this type of representation, a carbon atom is located at each corner of the six-sided ring. A hydrogen atom is attached to each carbon atom that does not have more than two carbon atoms. The condensed formula of KHP is $C_8H_5O_4K$.

For the standardization of NaOH solution, the equation below can be used to calculate the molarity of a sodium hydroxide solution. In that equation, M_{NaOH} is calculated when "m" grams of the potassium hydrogen phthalate (KHP), molar mass (MM_{KHP}), and "V_L" volume (in liters) of sodium hydroxide solution required are known.

$$m/MM_{KHP} = moles_{KHP} = moles_{NaOH} = V_L M_{NaOH}$$

Once standardized, the sodium hydroxide solution can then be used, if used within a short period of time, to titrate a different acid solution such as the carbonated beverage in this experiment.

To give a carbonated beverage a slight sour or tart taste, many manufacturers include an acid as one of the ingredients. The two acids most commonly used are citric acid ($HOC(CH_2CO_2H)_2CO_2H$), which has a condensed formula of $C_6H_8O_7$, and phosphoric acid (H_3PO_4) or a partially neutralized form of phosphoric acid such as NaH_2PO_4. Dark sodas such as the colas and root beer nearly always contain phosphoric acid since it is believed to enhance the taste of the caramel used to obtain the dark color. The less dark sodas, such as 7-Up™, usually have a more citrus, fruit-like taste due to the presence of citric acid as the acidic ingredient.

citric acid

TECHNIQUES

You will prepare solutions, use the balances, transfer solids and liquids, control a chemical reaction, use the burner, record observations, and make several calculations associated with an acid-base titration. The results of all calculations are to be expressed with significant figures only. These are skills and techniques that are considered to be essential for anyone who is preparing to enter any lab science career.

ACTIVITIES

You will prepare a sodium hydroxide solution of an approximate molarity, standardize the solution using potassium hydrogen phthalate, and use the standardized solution to determine, via titration, the number of acidic hydrogen atoms per molecule for citric acid and the concentration of citric acid in a carbonated beverage.

CAUTION	

Wear approved eye protection at all times in the laboratory. Sodium hydroxide is a strong base; handle it carefully and avoid contact with your skin. If contact has occurred, wash with plenty of tap water.

PROCEDURES

Preparation of the Soda Sample

1-1. Place in a 250-mL beaker about 90 mL of a clear soda. Determine the exact volume of the sample with a 100-mL graduated cylinder to the nearest 0.1 mL. Record the volume and identity of your sample.

1-2. Your sample may be carbonated. That is, it may contain a considerable amount of dissolved CO_2. The dissolved CO_2 will interfere with the citric acid determination. Therefore, you need to remove most of the dissolved carbon dioxide (Procedure 1-3) before you can analyze for citric acid. Before removing the CO_2, determine the initial acidity of the beverage.

Obtain a strip of pH Indicator Paper. To determine the pH of your sample, dip a stirring rod into the soda, and then touch the stirring rod to the indicator strip. Take the strip to the color chart posted by your instructor and compare the colors. Record your findings.

1-3. Heat your sample on a ring stand to near boiling. Allow it to cool. Remeasure its volume. Add distilled water to bring its volume back to its initial volume (Procedure 1-1). Determine its pH now by using a second indicator strip.

Preparation of 200 mL of 0.04 *M* NaOH

2-1. With the help of your instructor, recalculate the volume of 2.5 *M* sodium hydroxide solution needed to make 200 mL of a 0.04 *M* solution (question 5 on the Prelab).

2-2. Take a clean, but not necessarily dry, 10-mL graduated cylinder to the reagent area and obtain the volume of ~2.5 *M* sodium hydroxide calculated in Procedure 2-1. Pour the ~2.5 *M* NaOH solution obtained into a 500-mL Erlenmeyer flask containing about 100 mL of distilled water. Use 5 mL of distilled water to rinse the graduated cylinder. Add this wash to the flask. Add additional distilled water to the 500-mL Erlenmeyer flask until a total volume of 200 mL is reached. Mix thoroughly by carefully swirling. Stopper the flask and label it "~0.04 *M* NaOH."

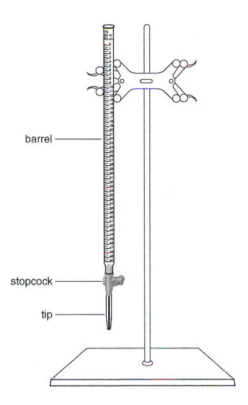

barrel ——

stopcock ——

tip ——

2-3. Set up a 50-mL buret. Check that it is clean, free of grease, and draining properly (see Common Procedures and Concepts Section at the end of this manual). Rinse the buret with three 5-mL portions of your ~0.04 *M* NaOH solution. Fill the buret with the NaOH solution to above the 0-mL mark. Drain some of the solution from the buret into a waste beaker to expel air from the buret tip. Allow the liquid level to drop below the zero mark. Restopper the flask containing the NaOH solution and set it out of the way.

Standardization of the 0.04 *M* NaOH Solution

3-1. In your PRELAB EXERCISES, you calculated the mass of potassium hydrogen phthalate needed to neutralize 15 mL of an ~0.04 *M* NaOH solution. Have your instructor check your calculations.

3-2. Tare a clean, creased piece of smooth paper (~4 inches × 4 inches) on the analytical balance. Carefully add to the paper approximately the mass of potassium hydrogen phthalate (KHP) needed to neutralize 15 mL of a 0.04 *M* NaOH solution. The mass actually obtained should not be exactly the amount you needed. *Remember that the balance must be clean when you leave it.* Record the mass of your sample. Transfer your sample of KHP to a clean but not necessarily dry 250-mL Erlenmeyer flask.

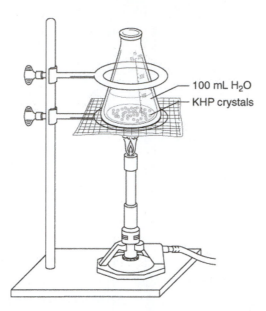

100 mL H$_2$O
KHP crystals

3-3. Add approximately 100 mL distilled water to the flask. If the KHP doesn't dissolve with swirling, gently warm and carefully swirl the flask until all the phthalate has dissolved. Then remove the flask from the flame and allow it to partially cool.

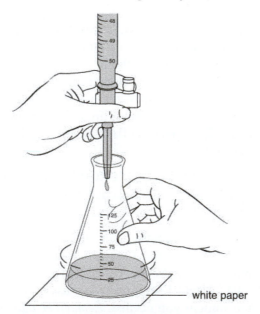

white paper

3-4. Add three or four drops of phenolphthalein indicator to the KHP solution in the flask and set the flask on a sheet of white paper under the buret. Read the volume in the buret to the nearest 0.01 mL. Record the reading of the volume. While gently swirling the warm flask, slowly add the ~0.04 M NaOH solution to the KHP solution until a pale pink endpoint is observed and that persists for about a minute. Record the final buret reading. Empty the flask and rinse it with distilled water. Repeat the standardization.

Titration of Citric Acid

4-1. Refill the buret with the NaOH solution, if needed. Tare a clean piece of smooth paper on the analytical balance. Obtain 0.04_ g of citric acid. Record the mass to the nearest milligram (0.001 g). Transfer the citric acid to a clean 250-mL Erlenmeyer flask.

4-2. Add approximately 100 mL distilled water to the flask. Carefully swirl and gently warm the flask until all of the solid has dissolved.

4-3. Add three or four drops of phenolphthalein indicator to the citric acid solution in the flask and set the flask on a sheet of white paper under the buret. Read the volume in the buret to the nearest 0.01 mL. Record the reading of the volume. While gently swirling the warm flask, slowly add the ~0.04 M NaOH solution until a pale pink endpoint is observed and persists for about a minute. Record the final buret reading. Empty the flask and rinse it with distilled water. Repeat the titration.

Titration of the Beverage

5-1. Check that your second 50-mL buret is clean, free of grease, and draining properly (see Common Procedures and Concepts Section at the end of this manual). Rinse the buret three times with 3 mL of your degassed sample of a beverage. Then fill the buret with the beverage sample to above the 0-mL mark. Drain some of the solution from the buret to expel air from the buret tip. Allow the liquid level to drop below the zero mark. Restopper the flask containing the remaining sample, label it, and set it aside.

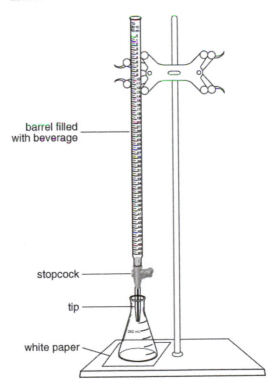

5-2. To the nearest 0.01 mL, read and record the volume level of your beverage sample in the buret. Slowly drain about 35 mL of the solution into a clean, but not necessarily dry 250-mL beaker. Record the final buret reading. Add four drops of phenolphthalein to the solution and titrate with your NaOH solution to a pale pink endpoint. Record the initial and final buret readings of the NaOH buret. Calculate the concentration of the citric acid in terms of both molarity and percent concentration by mass. If time allows, repeat the titration.

Clean-Up

6-1. All solutions may be safely flushed down the drain with large amounts of water.

6-2. All glassware must be rinsed and returned to their proper place.

6-3. After rinsing the burets, empty them and clamp them in the inverted position. Have your instructor sign your notebook and Report Form.

Name (Print) _____   Date (of Lab Meeting) _____   Instructor _____

Course/Section _____

10   EXPERIMENT 10: ANALYSIS OF A CARBONATED BEVERAGE

Prelab Exercises

1. Define:

   concentration (no equations, please)

   acid

   endpoint

   standardization of a solution

2. Calculate the molar mass of citric acid.

3. Sometimes there is an air bubble in the stopcock of the buret. How would the air bubble affect your measurements?

4. How does one remove all air bubbles from a buret?

5. Calculate the volume of ~2.5 *M* NaOH solution needed to prepare 200 mL (2 significant figures) of a ~0.040 *M* NaOH solution. (Record your answer here and in your notebook.)

6. Calculate the mass of solid sodium hydroxide needed to prepare 200 mL (2 significant figures) of a 0.040 *M* solution. If one pellet of sodium hydroxide weighs 0.170 g, how many pellets should be dissolved?

7. Draw a structural representation of KHP that shows all atoms. Then give two more representations that illustrate different extents to which formulas can be condensed or expanded.

8. Write a balanced equation for the neutralization of potassium hydrogen phthalate ($C_8H_5O_4K$) with sodium hydroxide. Calculate the mass of potassium hydrogen phthalate that neutralizes 15 mL of an ~0.040 *M* NaOH solution. (Record your answer here and in your notebook.)

Date _____   **Student's Signature** _____

_____ _____ _____
Name (Print)   Date (of Lab Meeting)   Instructor

_____ _____
Course/Section   Partner's Name (If Applicable)

10   EXPERIMENT 10: ANALYSIS OF A CARBONATED BEVERAGE

Report Form

DATA

Beverage Sample

Volume of sample: _____ mL   Name of beverage: _____

Initial pH of sample: _____   pH after degassing: _____

Solution Preparation

Concentration of stock NaOH solution used: _____   Volume of stock NaOH soln. used: _____

Mass of KHP calculated to neutralize 15 mL of ~0.04 M NaOH: _____ g

Solution Standardization

	Trial 1	Trial 2		Trial 1	Trial 2
phthalate _____ g		_____ g	Final volume NaOH _____ mL		_____ mL
			Initial volume NaOH _____ mL		_____ mL
			Vol. NaOH consumed _____ mL		_____ mL

Determination of Citric Acid

	Trial 1	Trial 2		Trial 1	Trial 2
citric acid _____ mL		_____ mL	Final volume NaOH _____ mL		_____ mL
			Initial volume NaOH _____ mL		_____ mL
			Vol. NaOH consumed _____ mL		_____ mL

Determination of Beverage Sample

	Trial 1	Trial 2		Trial 1	Trial 2
Beverage _____ mL		_____ mL	Final volume NaOH _____ mL		_____ mL
			Initial volume NaOH _____ mL		_____ mL
			Vol. NaOH consumed _____ mL		_____ mL

Date _____   **Instructor's Signature** _____

ANALYSIS

Solution Standardization

	Trial 1	Trial 2
Standardized concentration of NaOH solution:	_____ *M*	_____ *M*

Average standardized concentration of NaOH solution: _____ *M*

Determination of Citric Acid

	Trial 1	Trial 2
Moles of citric acid:	_____ moles	_____ moles
Moles of citric acid:	_____ moles	_____ moles

Relationship between Moles of Citric Acid and Moles of NaOH

Number of moles of NaOH per molecule of citric acid: _____

Number of moles of citric acid per mole of NaOH: _____

Proposed equation for the reaction between NaOH and citric acid:

Number of acidic hydrogen atoms per molecule of citric acid: _____

Number of molecules of citric acid per acidic hydrogen atom: _____

Determination of Beverage Sample

	Trial 1	Trial 2
Moles of NaOH consumed:	_____ moles	_____ moles
Moles of citric acid neutralized:	_____ moles	_____ moles
Molarity of citric acid in beverage sample:	_____ *M*	_____ *M*
Average *M* of beverage:	_____ *M*	
% Concentration by mass: (assume a density of 1.00 g/mL)	_____ %	_____ %
Average % concentration:	_____ %	

POSTLAB QUESTIONS

1. What are three of the properties that one looks for when selecting a primary standard?

2. Why is a solution of NaOH that was standardized yesterday unsuitable to be used as a standard solution today?

3. If KHP is a monoprotic acid, how would you describe citric acid? Explain your reasoning.

4. Draw a particle view of the reaction between NaOH and citric acid.

5. Describe the meaning of each in terms of its application to acid/base titrations.

   (a) standardization

   (b) indicator

   (c) endpoint

6. Describe three techniques used in this experiment.

7. Could this experiment be modified to work with a dark soda? If so, how?

Date _____ **Student's Signature** _____

Mass Relationships in Reactions

An Open Inquiry Experiment

INTRODUCTION

You have completed a number of labs dealing with the stoichiometry or mass relationships between substances in reactions. In prior experiments, the manual has specified the data to be collected and the chemicals to be used; however, in this experiment you will choose your chemical system and design the experimental procedures you wish to use. This type of experiment is an open inquiry lab. "Open" means that you choose the chemical system you wish to investigate from a number of options and you design the procedures. You will want to collect variables that provide evidence of the mass relationships in the reaction(s) you choose to investigate.

OBJECTIVES

You (with the assistance of a partner) will design an experiment that will allow you to collect data on the mass relationships for certain reaction(s). The details of your experimental design will depend on the question you choose to investigate.

CONCEPTS

This experiment uses the concepts of mass, volume, solution preparation, and stoichiometry.

TECHNIQUES

Graphing experimental data and transferring liquids and solids are just some of the techniques encountered in this experiment.

ACTIVITIES

You will design an experiment that will provide information about the mass relationships among the substances in a chemical reaction.

CAUTION	

You will be working with dilute solutions but you should treat all solutions as potentially dangerous. Dispose of materials as directed by your instructor. Wear approved eye protection.

PROCEDURES

Experimental Design

1-1. Design an experiment from the options in 1-3.

The design must include:

Problem Statement: This includes a few sentences describing the specific question(s) you are trying to answer with your experiment and propose equations for the reactions that you will use.

Proposed Procedures: This section contains the materials and equipment that you will use, the type of data you will collect (the variables you will measure), and the number of trials you are proposing. Remember to discuss safety considerations. Your planned experimentation should take up ⅔ of the lab period.

1-2. Remember that multiple runs will establish the reliability of your data. Try to keep the amounts small (usually < 1 g) and the concentrations dilute (~0.5 to .1 M).

1-3. Choose from the following six options:

Option 1

Investigate the relationships between the masses of reactant and product(s) associated with heating various compounds. Choose one of the following:

magnesium carbonate—$MgCO_3$

magnesium—Mg

potassium bicarbonate—$KHCO_3$

Option 2

Investigate the relationships associated with the mass changes on heating of a hydrated compound. Choose one of the following:

hydrated iron(II) sulfate
hydrated sodium thiosulfate

Option 3

Investigate the mass relationships associated with adding acids or bases to salts. Choose one of the following:

2.5 *M* sodium hydroxide added to sodium hydrogen phosphate, sodium dihydrogen phosphate, and sodium phosphate

5 *M* hydrochloric acid added to potassium carbonate and potassium bicarbonate

Option 4

Investigate the relationships associated with the formation of percipitates when mixing solutions of ionic compounds. Choose one of the following:

calcium chloride and sodium carbonate

copper(II) sulfate and sodium carbonate

copper(II) sulfate and sodium phosphate

iron(III) nitrate and sodium phosphate

Option 5

Investigate the relationships associated with the addition of metals to hydrochloric acid. Choose one or more of the following:

iron

magnesium

zinc

Option 6

Modify any of the above or investigate a mass relationship of your choosing.

1-4. Discuss your ideas with your instructor at least a few days prior to the lab to determine chemical and equipment availability. Your instructor will supply feedback on your experimental design BEFORE you do the experiment.

Experimentation   **2-1.** Make any changes that your instructor suggests, and then proceed to collect data. You will collect data with a partner and should discuss your results with your partner. These discussions will help you both learn.

2-2. Record the data in your notebook and get your instructor's signature when all data is collected.

Report

3-1. After completing the data collection, you will write a lab report. Although you collect data and share ideas with a partner, you will be expected to write the final lab report independently. Your grade will depend on the thoroughness of your investigation, the presentation of your data, the careful analysis of the data, and the logic put in to give reasonable results and explanations.

3-2. The lab report MUST include the following four sections:

Problem Statement: This includes a few sentences describing what specific questions you were trying to answer with your experiment and proposed equations for all reactions that were used.

Procedures: This section contains the materials and equipment that you actually used (these may differ from those you proposed), the type of data collected (the variables measured), and the number of trials done. Remember to discuss safety considerations. Your experimentation should take up ⅔ of the lab period.

Data/Analysis: Include the data you collected. Data should be in tables when possible, with easy-to-read labels. Analysis of the data should also be included (analysis = what does your data tell you). Graphs (with labels, units, and titles), mathematical relationships, chemical equations, and algebraic equations should be given, and the connection to the data should be shown.

Conclusion: This is the generalization or explanation you have deduced from your experiment. This is also the place to make explanations for any data results that are counter to logical chemical ideas and to describe how you would change the experiment if you repeated it.

11   EXPERIMENT 11: MASS RELATIONSHIPS IN REACTIONS

Prelab Exercises

Problem Statement

Proposed Procedures (insert more sheets if needed)

Date _____   **Student's Signature** _____

Instructor's Approval and Comments:

Date _____ **Instructor's Signature** _____

11   EXPERIMENT 11: MASS RELATIONSHIPS IN REACTIONS

Report Form

DATA

Data is collected in your notebook.

Date _____   **Student's Signature** _____

LAB REPORT

Attach this sheet to your lab report that includes the PROBLEM STATEMENT (actual), PROCEDURES (actual), DATA/ANALYSIS, and CONCLUSION.

POSTLAB QUESTIONS

1. The title of this experiment is Mass Relationships in Reactions. What is the difference between one mole of two different compounds? What mass unit is used in chemical equations or reactions?

2. What are the most important factors to keep in mind when designing an experiment? Discuss at least three.

3. Draw a particle view(s) of the reaction(s) in your experiment.

4. Answer any questions assigned by your instructor.

Date _____ **Student's Signature** _____

Shapes of Molecules and Ions

A Guided Inquiry Experiment

INTRODUCTION

To write the Lewis structure of a compound or polyatomic ion, follow these steps:

1. Select a reasonable (symmetrical) skeleton for the molecule or polyatomic ion.

    a. The least electronegative element is usually the central atom; however, hydrogen is never the central atom.

    b. The central atom is also usually the one needing the most electrons to fill its octet.

    c. Oxygen atoms do not bond to other oxygen atoms except in O_3, O_2, and peroxides such as H_2O_2.

    d. The skeleton is also referred to as the spider.

2. Count the valance electrons present.

3. Try to fill in the valance electrons in such a way that only single bonds and nonbonding valance electrons are present.

4. Check for compliance with the "octet rule."

    a. If some atoms have less than the octet of electrons, consider the possibility of multiple bonds. Remember that most elements do not form multiple bonds.

    b. If there are more electrons than needed for single bonds and non-bonding electrons to satisfy the octet rule, consider expanding the octet on the central atom.

5. Adjust the drawing to account for bonding knowledge, resonance, etc.

To determine the geometry of the compound or ion, follow these steps:

1. Follow the steps above to obtain a correct Lewis structure.

2. Count the regions of high electron density "on" the central atom.

3. Determine the electronic or base geometry by giving each region of high electron density its maximum space.

4. Determine the molecular (or actual) geometry by describing the shape created within the electronic geometry by only the bonded atoms (consult the table in the Common Procedures and Concepts Section at the end of this manual for additional information).

5. Adjust the molecular geometry to account for the space requirement and electronic groups of different sizes.

To determine if a compound is polar, follow these steps:

1. Follow the steps above to obtain a correct Lewis structure with a known geometry.

2. Ask the following two questions.

   a. Are there polar bonds that are not arranged so that they cancel?

   b. Are there lone pairs on the central atom that are not arranged so that they cancel?

3. If the answer to either question is yes, the compound is polar.

OBJECTIVES

In this experiment, you will look for patterns as you practice writing Lewis structures for molecules and ions, predicting their shapes, and stating whether or not the molecules are polar.

CONCEPTS

This experiment uses the concepts of valence electrons, Valence Shell Electron Repulsion Theory (VSEPR), electronegativity, hybridization, polar bonds, formal charges, and polar molecules.

ACTIVITIES

You will be assigned a number of molecules and/or ions for which you will be asked to determine the number of valence electrons, the Lewis structure, formal charge, geometry, 3-D model, bond angles, polarity, hybridization, and number of pi bonds.

CAUTION

Use approved eye protection if anyone in the lab is using any glassware or chemicals.

PROCEDURES

1-1. You will be assigned one molecule or ion from each of the following groups.

Possible Polyatomic Ions and Molecules

List #1	List #2	List #3	List #4
1. $SiCl_4$	1. CH_3Br	1. PO_4^{3-}	1. NH_4^+
2. PCl_3	2. ICl_2^+	2. H_2NNH_2	2. SF_2
3. NO_3^-	3. NO_2^-	3. CS_2	3. COF_2
4. SF_6	4. BBr_3	4. BCl_3	4. SO_2
5. PCl_5	5. $HOOH$	5. SbF_5	5. PBr_5
6. IF_3	6. $SbCl_5$	6. PCl_6^-	6. $TeCl_4$
7. XeF_5^+	7. BrF_3	7. ClF_5	7. XeF_4
8. SO_4^{2-}	8. ICl_4^-	8. XeF_2	8. ClO_4^-

1-2. You will be assigned a fifth compound that may not be included in the lists above. A sixth compound may also be assigned.

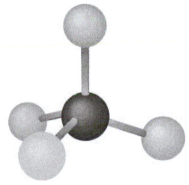

1-3. Draw the Lewis structure of each of the compounds assigned. Show any net formal charge. Additionally, your instructor may ask you to show all formal charges. For more on Lewis structures, consult your textbook.

1-4. Make a model of each. Determine the molecular geometry of each. Draw a 3-D representation of your models.

1-5. Label the bond angles. Determine the polarity of the bonds and the molecules.

1-6. Determine the hybridization of the central atom and the presence or absence of pi bonds.

1-7. Have your instructor approve your model and sign your notebook and Report Form.

Name (Print)   Date (of Lab Meeting)   Instructor

Course/Section

12   EXPERIMENT 12: SHAPES OF MOLECULES AND IONS

Prelab Exercises

1. Describe what is meant by "the octet rule."

2. What is meant by an octet rule violation? Give two examples of how a molecule could violate the octet rule and give a molecule that violates the rule in this fashion.

3. Draw a 3-D representation of a compound that is trigonal pyramidal.

4.   When can a compound containing polar bonds become a nonploar molecule?

Date _____   **Student's Signature** _____

| 12 | **EXPERIMENT 12: SHAPES OF MOLECULES AND IONS** |

Report Form

DATA/ANALYSIS

Give the following for each ion or molecule assigned to you:

Assigned ion or molecule _____	**Assigned ion or molecule** _____
1. Number of valence electrons _____	1. Number of valence electrons _____
2. Attach Lewis structure including any net formal charges.	2. Attach Lewis structure including any net formal charges.
3. Molecular geometry:	3. Molecular geometry:
4. 3-D model approved _____ Attach 3-D sketch	4. 3-D model approved _____ Attach 3-D sketch
5. Show the bond angles on attached Lewis structure.	5. Show the bond angles on attached Lewis structure.
6. If the assigned compound is a molecule, is it polar? Explain:	6. If the assigned compound is a molecule, is it polar? Explain:
7. Hybridization of the central atom _____	7. Hybridization of the central atom _____
8. Number of pi bonds present _____	8. Number of pi bonds present _____

145

Assigned ion or molecule _____

1. Number of valence electrons _____
2. Attach Lewis structure including any net formal charges.
3. Molecular geometry:

4. 3-D model approved _____
   Attach 3-D sketch
5. Show the bond angles on attached Lewis structure.
6. If the assigned compound is a molecule, is it polar? Explain:

7. Hybridization of the central atom _____

8. Number of pi bonds present _____

Assigned ion or molecule _____

1. Number of valence electrons _____
2. Attach Lewis structure including any net formal charges.
3. Molecular geometry:

4. 3-D model approved _____
   Attach 3-D sketch
5. Show the bond angles on attached Lewis structure.
6. If the assigned compound is a molecule, is it polar? Explain:

7. Hybridization of the central atom _____

8. Number of pi bonds present _____

Assigned ion or molecule _____

1. Number of valence electrons _____
2. Attach Lewis structure including any net formal charges.
3. Molecular geometry:

4. 3-D model approved _____
   Attach 3-D sketch
5. Show the bond angles on attached Lewis structure.
6. If the assigned compound is a molecule, is it polar? Explain:

7. Hybridization of the central atom _____

8. Number of pi bonds present _____

Assigned ion or molecule _____

1. Number of valence electrons _____
2. Attach Lewis structure including any net formal charges.
3. Molecular geometry:

4. 3-D model approved _____
   Attach 3-D sketch
5. Show the bond angles on attached Lewis structure.
6. If the assigned compound is a molecule, is it polar? Explain:

7. Hybridization of the central atom _____

8. Number of pi bonds present _____

Date _____   **Instructor's Signature** _____

POSTLAB QUESTIONS

1. Why were you not asked to predict whether or not an ion is polar?

2. Explain the hybridization of the bonding orbitals of carbon in formaldehyde (H_2CO).

3. Explain the hybridization of the bonding orbitals of sulfur in the SO_3^{2-} ion.

4. Certain geometries tend to be present in polar molecules. List four. What do these geometries have in common?

5. Is it possible for the central atom to have more than 2 pi bonds? Explain your reasoning.

6. Why is hydrogen never the central atom in a Lewis structure?

7. Give two examples in which the predicted bond angles are different from the "actual bond angles."

8. Explain how molecular shape is predicted. What is the determining factor?

9. Explain how the electronic geometry may be the same as the molecular geometry or shape of the molecule.

10. Draw a 3-D representation of a compound that has a seesaw shape.

Date _____ **Student's Signature** _____

The Fuel in a Bic® Lighter

A Guided Inquiry Experiment

INTRODUCTION

When chemists prepare new compounds, they must characterize the compounds by determining such properties as melting point, boiling point, color, one or more types of spectral analysis, and/or elemental composition. In this experiment, you will investigate an unknown volatile hydrocarbon.

OBJECTIVES

You will use data obtained in lab to relate the mass, volume, temperature, and pressure of a gas.

CONCEPTS

The ideal gas equation, general gas laws, pressure, partial pressure, volume, standard conditions, and the standard molar volume are several of the concepts used during this experiment.

TECHNIQUES

You will displace water in an inverted container with a sample of gas released from a disposable lighter. The method used is applicable to substances that are gases at room temperature or can be easily volatilized by heating and are not soluble in water. Your understanding of properties of gases will be used throughout this experiment. You will also make extensive use of the top-loader balance.

ACTIVITIES

You are to investigate the hydrocarbon that is the liquid you see in a disposable lighter. You are to weigh a lighter on the top-loader balance, hold the lighter under an inverted container filled with water, and allow the fuel to displace part of the water in the inverted container. By knowing the volume change in the water in the container, the temperature and partial pressure of the water, the temperature of the collected gas, and the room (or total) pressure, you will be able to calculate the volume that the collected gas would occupy when dry and at standard conditions. Then you are to study the relationship between calculated volume of gas and the change in mass of the lighter.

> **CAUTION**
>
> You are working with volatile hydrocarbons under pressure. Do not allow open flames to come in contact with the gaseous hydrocarbon. Wear approved eye protection while in the laboratory.

PROCEDURES

1-1. Fill a large beaker, plastic tray, or other large container with tap water. Fill a 100-mL graduated cylinder with distilled water. Place the palm of one hand over the top of the filled cylinder, invert it, and place it in the beaker or tray of water. Remove your hand. There should be no bubble (or only a very small bubble) at the top of the cylinder.

1-2. Take a disposable lighter to the analytical balance. Tare the balance with a small sheet of paper on it. Weigh the disposable lighter to the nearest milligram. Record its mass in your notebook. Hold the lighter by the small sheet of paper to avoid excess handling of the lighter.

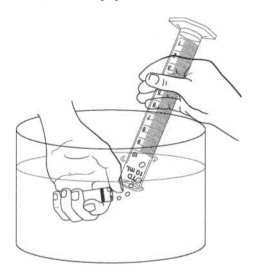

1-3. Take the disposable lighter out of the sheet of paper and hold it under the mouth of the inverted cylinder. Depress the lever on the lighter so that fuel escapes from the lighter but is captured in the inverted cylinder. Continue until the gas in the inverted cylinder is about an inch from its top. You will need to stop within the range of the volume markings on the cylinder. Remove the lighter and lay it on a dry towel or cloth.

1-4. Hold the cylinder so that it is vertical and the levels of the water inside and outside are equal. Mark the level with your finger. Remove the cylinder. Determine the volume of the hydrocarbon that you just released.

1-5. Wipe and shake as much water as practical from the disposable lighter without causing any more fuel to escape. Take the lighter back to the top-loader balance. Re-tare a small sheet of paper and weigh the lighter.

1-6. Repeat Procedures 1-1 through 1-5 until you have four sets of data that you believe to be reliable. Do not count your first try at collecting gas. Vary the volume during each set of data; range from 50 to 100 mL of gas in your graduated cylinder. (Your first set of data will probably not be consistent with the other sets of data.)

1-7. Record the temperature of your water bath and the barometric pressure. Knowing the temperature of the water, you can go to tables such as the one below and determine the partial pressure due to water vapor. The barometric pressure at the time of the experiment will be equal to the partial pressure of the hydrocarbon plus the partial pressure of the water vapor (if the water levels inside and outside the container were the same each time the volume was noted).

Vapor Pressure of Water

°C	16	17	18	19	20	21	22	23	24	25	26	27	28	29	30	31
Torr	13.6	14.5	15.5	16.5	17.5	18.7	19.8	21.1	22.4	23.8	25.2	26.7	28.3	30.0	31.8	33.7

1-8. Have your instructor sign your notebook and Report Form. Using your data set, complete the Report Form.

Name (Print) | Date (of Lab Meeting) | Instructor

Course/Section

13

EXPERIMENT 13: FUEL IN A BIC® LIGHTER

Prelab Exercises

1. Define:

atmospheric pressure

standard temperature and pressure

vapor pressure of water

Kelvin temperature scale

2. What is the name that we give to the equation, PV = nRT? (Define each term.)

3. In this experiment, what measurement will limit the number of significant digits in your final calculation?

4. What can be done physically to assure that the total pressure of the system is the same as the current atmospheric pressure?

5. Draw a particle view of the gases that are present in a container in which hydrogen gas is being collected over water.

Date _____ **Student's Signature** _____

13 EXPERIMENT 13: FUEL IN A BIC® LIGHTER

Report Form

DATA

Data is collected in your notebook.

Water bath temperature: _____ °C _____ K

Barometric pressure: _____ torr _____ atm

Vapor pressure of water at the water bath temperature: _____ torr

Partial pressure of hydrocarbon: _____ torr _____ atm

	Experimental Trials			
	1	2	3	4
Initial Mass of Lighter				
Final Mass of Lighter				
Mass of Hydrocarbon				
Volume of Hydrocarbon				

Date _____ **Instructor's Signature** _____

ANALYSIS

	Experimental Trials			
	1	2	3	4
Vol of Hydrocarbon at STP				
Mass of Hydrocarbon				
Mass of Hydrocarbon + Vol				
Mass of Hydrocarbon – Vol				
Mass of Hydrocarbon × Vol				
Mass of Hydrocarbon ÷ Vol				

Sample calculation for one trial:

1. What pattern do you find between the volume and mass data? Look at the different mathematical operations (+, –, ×, ÷) above. Do any of the operations yield a value that is nearly constant? If you found one or more of the operations to be a constant, what is(are) the name(s) for the constant(s) you found? What are its units?

2. Use the ideal gas equation to derive the volume of one mole of an ideal gas at STP (this is called the standard molar volume).

3. Explain how you could use a set of values like those you obtained in question 1 and the standard molar volume to find the grams/mole of the fuel. (HINT: Look closely at the units.)

4. Using the method you described in question 3, find the grams/mole for each of your experimental trials.

5. What is another name for grams/mole? For the values in question 4, find the average value, the standard deviation, and the relative standard deviation.

6. Give the ideal gas equation ($PV = nRT$). Substitute g/molar mass for the moles of gas in the ideal gas equation and rearrange the equation to isolate molar mass.

7. For each of your four experimental trials, identify the variables in the above equation for molar mass. Calculate the value obtained for molar mass for each of your four trials.

8. How does the answer to question 7 compare with the value you found in question 4? Explain your answer.

9. Given that the formula of butane is C_4H_{10}, does you answer to question 4 approximately equal the molar mass (MM) of butane? Is your value closer to that of propane?

10. What is the accuracy of your findings? 100 (theoretical – experimental MM): _____ Find the % deviation of your MM from the accepted MM of butane.

11. Based on your answer to question 9 above, which other hydrocarbons listed below could be mixed with butane to give the average molar mass you determined?

Formulas of Small Hydrocarbons	*Names*
CH_4	methane
C_2H_6	ethane
C_3H_8	propane
C_4H_{10}	butane
C_5H_{12}	pentane
C_6H_{14}	hexane
C_7H_{16}	heptane
C_8H_{18}	octane

POSTLAB QUESTIONS

1. Except for very small alkanes (hydrocarbons), the boiling point rises 20° to 30° for each additional carbon atom in the molecule. Assume that the normal boiling point of the fuel in the lighter is 10°C, why was it not necessary to extend the table further (i.e., why was it unlikely that your unknown contained more than eight carbon atoms per molecule)?

2. How is the ideal gas law related to the molar mass of a gas?

3. A gaseous hydrocarbon collected over water at a temperature of 23°C and a barometric pressure of 753 torr occupied a volume of 50.5 mL. The hydrocarbon in this volume weighs 0.1433 g. Calculate the molecular mass of the hydrocarbon. What is the identity of the hydrocarbon?

Date _____   **Student's Signature** _____

Alka-Seltzer®: An Application of Gas Laws

A Guided Inquiry Experiment

INTRODUCTION

Nearly everyone in the United States and in many other countries is familiar with the popular antacid Alka-Seltzer®. It is well known for its fizzing when placed in water. It "fizzes" when dropped into water because the dissolved sodium bicarbonate, also known as baking soda or sodium hydrogen carbonate, reacts with the citric acid that is also present to liberate gaseous carbon dioxide. You should be able to write the equation for this reaction. This reaction does not occur until the reactants are dissolved in water. In this experiment, you will investigate the reaction of the sodium hydrogen carbonate in the Alka-Seltzer tablets with hydrochloric acid.

OBJECTIVES

During this experiment, you are to apply several different gas laws to the carbon dioxide produced by a commercial product. You will gain experience with Boyle's, Charles's, Dalton's, Avogadro's, Gay-Lussac's, and Henry's Laws. You will examine the relationship between the gas laws and the stoichiometry of balanced equations. You are to use an apparatus similar to that described below and you are to design an experiment that will provide the data needed for the requested stoichiometric calculations. From the volume of carbon dioxide that you determine to be liberated from your portion of an Alka-Seltzer tablet, you are asked to give evidence for your proposed equation and for the percent hydrogen carbonate present in a whole tablet.

CONCEPTS

This experiment uses the following stoichiometric concepts: (1) Moles of hydrogen carbonate is related to the mass of the tablet. Mass of hydrogen carbonate in whole tablet = (mass of whole tablet/mass of sample) × mass determined by experimentation. (2) Moles of CO_2 are obtained from the volume of CO_2 produced plus the volume of CO_2 dissolved in the acid solution, both corrected to standard temperature and pressure before conversion to moles.

Research scientists often find procedures in the literature for similar reactions to those they are investigating. They must decide whether to use a modified procedure or to use the same procedure. The most important thing is that they must be able to collect the data they need. You also have specific data that you will need to collect. Assume that you have found the following procedure in the literature. You must decide the best way to collect your data.

Preparing the Apparatus

Position a buret clamp about 30 cm above the base of a ringstand so that the arm of the clamp extends over the base. Near the top of the ringstand attach a small split-ring clamp at approximately 45° angle to the first clamp. Also near the top of the ringstand, position a clamp that is suitable for holding a 125- mL Erlenmeyer flask. Attach a 60-cm-long piece of rubber tubing to the lower end of a 50-mL buret. Clamp the buret securely in the buret clamp. The lower end of the buret should be about 15 cm above the desktop. Obtain a No. 00 one-hole rubber stopper that holds a short piece of glass tube with a 90° or smaller bend and a one-hole stopper that will fit (stopper) a 125-mL Erlenmeyer flask. There must be a short piece of straight glass tube in the stopper. Fasten the 125-mL flask securely to the ringstand. Connect the flask and the top of the buret with the 60-cm-long rubber tubing.

Position a leveling bulb in the split ring on the ringstand. Connect the rubber tubing attached to the bottom of the buret to a leveling bulb.

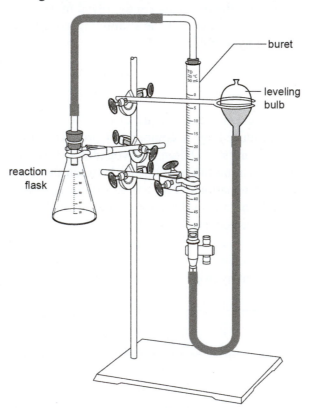

The apparatus shown may be used in the following way to collect data.

Preparing Tablet Fragments

Obtain an Alka-Seltzer tablet or a tablet of a similar preparation. To keep the tablet dry and protected from moisture it should be *kept wrapped in aluminum foil or plastic wrap whenever you are not breaking off a sample.*

Determine the mass of the whole tablet on the analytical balance. Quickly place the tablet back in the foil wrapper. Practice breaking off approximately one tenth of the tablet. Quickly determine the mass of the fragment and return it to the foil wrapper. Use this result to estimate the size of a fragment weighing 0.2 to 0.3 g. Such a fragment will be needed later.

Preparing ~6 *M* HCl Solution

If it is not provided, prepare 25 mL of ~6 *M* hydrochloric acid by filling a graduated cylinder to the 12.5-mL mark with distilled water and ice. While stirring, slowly add concentrated (12 *M*) HCl until the solution reaches the 25-mL mark. Stir thoroughly, but carefully; then remove the stirring rod and read (to the nearest 0.5 mL) and record the volume of the liquid in the cylinder. Allow this solution to cool to room temperature.

Preparing a Saturated CO_2 Solution

To prepare an aqueous solution saturated with carbon dioxide, place 125 mL of distilled water in a 250-mL or 400-mL beaker and carefully add 4.0 mL of 6 *M* hydrochloric acid. Stir the mixture. Drop half of the tablet including all small fragments into the acidic solution. Stir the mixture until bubbling has ceased. Solid materials may remain.

Filling the Apparatus

Decant the solution that is saturated with carbon dioxide into the leveling bulb. Loosen the stopper in the Erlenmeyer flask. Lift the leveling bulb and slide the rubber tubing through the opening in the ring. Raise and lower the leveling bulb until there appears to be no more trapped gas bubbles in the rubber tubing. Place the bulb back into the ring.

Record (to the nearest 0.5 mL) the volume of ~6 *M* acid solution in your graduated cylinder. Place about 5 mL of the acid solution in the Erlenmeyer flask. Read and record the level of the acid remaining in the cylinder. If necessary, dry the inside neck of the flask and the stopper with a towel. Restopper the flask.

Running the Experiment

Plan the next steps carefully and perform them quickly. Take the remaining piece of your Alka-Seltzer tablet that is wrapped in foil to the analytical balance. Quickly break off a piece estimated to weigh 0.2 to 0.3 g. Wrap the remaining part of the tablet with the foil. Quickly determine the mass of the piece of tablet. If the piece weighs more than 0.3 g, chip off a small piece and reweigh. Repeat this procedure until the mass of the piece is between 0.2 and 0.3 g. Record its mass to the nearest 0.1 mg (0.0001 g).

Using a thin thread about 20-cm long, quickly form a loop, slip the weighed tablet piece into the loop, and pull it snug.

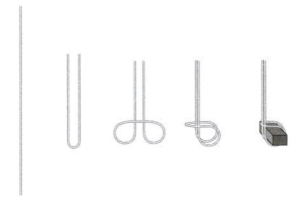

Unstopper the Erlenmeyer flask and raise the leveling bulb until the liquid level in the buret is between the 0-mL and 2-mL mark.

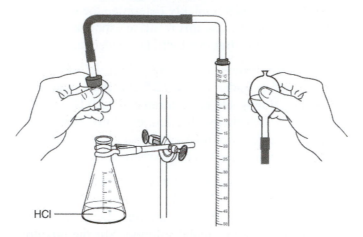

Then carefully suspend the tablet sample about 2 cm above the acid solution in the Erlenmeyer flask. Hold the thread in place and slowly push the stopper snugly into the neck of the flask. Be sure the stopper fits tightly.

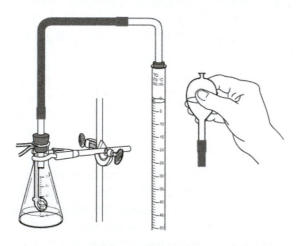

CAUTION

Do not exert too much pressure on the stopper. You might break the neck of the flask and cut yourself.

Lower the leveling bulb to a position ~10 cm below the ring clamp. Watch the water level in the buret. The level should drop and come to rest above the level in the bulb. When the liquid level in the buret does not come to rest and slowly descends, the system is leaking. Check both stoppers for leaks. Carefully tighten the two stoppers with a twisting motion and some downward pressure.

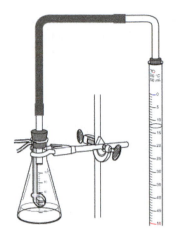
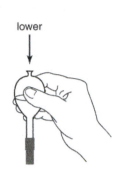

lower

Repeat the test for leaks. If you continue to have a leak, moisten the connections or consult your instructor. (Slow leaks may be okay provided that you never allow the levels of the surfaces of the liquids in the leveling bulb and in the buret to become greatly different.)

When your system has passed the leak test, raise the bulb until the liquid level in the buret and the liquid level in the bulb are at the same height. Under these conditions the liquid level in the buret should be between the 1-mL and 5-mL marks. When the buret reading is in the correct range, record the reading to the nearest 0.01 mL.

After you have read the liquid level in the buret with the level in the bulb at the same height, replace the bulb in the ring. The critical "fast steps" are now finished.

Carefully unclamp the Erlenmeyer flask and tilt it until the acid contacts the suspended piece of Alka-Seltzer. The sample will soon slip out of the thread. Be sure it has dropped into the solution; then reclamp the flask loosely in position to permit you to shake the flask gently until the CO_2 evolution has ceased.

Keep the liquid levels in the buret and bulb at nearly the same height during gas evolution by lowering the bulb at the same rate at which the level in the buret drops. This precaution will minimize errors caused by leaks.

Continue careful checking until the liquid level in the buret has stabilized. Then adjust the leveling bulb to match liquid levels and read and record the liquid volume in the buret to 0.01 mL.

Unstopper the Erlenmeyer flask, discard the solution in the flask, rinse the flask with tap water and then with distilled water, and dry the inside of the flask with a towel. Repeat the experiment with new pieces of the Alka-Seltzer tablet until you have a suitable number. When you have completed three additional satisfactory runs of the experiment, clean the Erlenmeyer flask, discard the liquid in the buret and bulb, and rinse all glassware with distilled water. Your instructor will let you know what you should do with the apparatus.

TECHNIQUES

Preparation of an acid solution, assembly of a glass apparatus, and determination of mass, temperature, and pressure are just some of the techniques encountered in this experiment.

ACTIVITIES

You will measure the volume of carbon dioxide produced by a portion of an Alka-Seltzer tablet. Knowing the chemistry that takes place, its volume, the temperature of the water it was collected over, the pressure at which it was collected, and the amount that will be dissolved in the acidic solution, you are to relate the data to the proposed equation and to the percent bicarbonate ion in the tablet.

CAUTION

You will be handling glass equipment and working with acidic solutions. Common sense will help one to avoid injuries to oneself or to other students or the damage of property. Wear eye protection at all times. Under no circumstances should the reaction of Alka-Seltzer (or a similar preparation) with acid or water be performed in a closed container. Be very careful when working with glassware, particularly when tightening stoppers and slipping rubber tubing on glass tubes. Excessive force in these operations might lead to breakage of the glass and painful cuts.

PROCEDURES

Constructing Equipment

1-1. Construct the equipment described in the Concepts section above.

Running of Experiment

2-1. Use the equipment described in the Concepts section above to obtain four sets of data for the mass of a tablet piece, the volume of CO_2 collected, the barometric pressure, and the temperature in the laboratory.

2-2. Your task is to relate your experimental data (grams, moles, volume) to your proposed equation for the reaction of HCl with $NaHCO_3$ and percentage of HCO_3^- in a tablet.

Calculations

3-1. Calculate the partial pressure of carbon dioxide in each of your experimental runs. You will need to apply Dalton's Law of Partial Pressures. You will need the temperature of the acid solution in the flask, of the carbon dioxide saturated solution in the leveling bulb, and of the gas obtained. These normally are at room temperature, which will make the calculations easier. Use the temperature and the table below to determine the partial pressure due to water vapor. The total pressure of the gaseous sample will be the partial pressure of the carbon dioxide plus the partial pressure of the water vapor.

Vapor Pressure of Water

°C	16	17	18	19	20	21	22	23	24	25	26	27	28	29	30	31
Torr	13.6	14.5	15.5	16.5	17.5	18.7	19.8	21.1	22.4	23.8	25.2	26.7	28.3	30.0	31.8	33.7

3-2. Calculate the volume, which the carbon dioxide would occupy at the pressure of 1 atm (760 torr) and 25°C. You will need to add the volume of carbon dioxide that dissolved in the hydrochloric acid in the Erlenmeyer flask. The solution in the leveling bulb was already saturated with carbon dioxide so no new carbon dioxide should have dissolved in that solution.

The amount of carbon dioxide that will dissolve in the acid solution in the flask is a function of pressure (Henry's Law) and temperature. Experiments beyond our laboratory capabilities showed that 0.80 mL of gaseous carbon dioxide at 760 torr and 25°C will dissolve in each mL of 6.0 *M* HCl. (The value varies with temperature and pressure but over small ranges of temperature and pressure the variations are known to be small and in many cases the two factors vary in such a way that they tend to counteract each other. For this experiment, we will treat the solubility of carbon dioxide in the small volume of acid as a constant.)

The volume which the carbon dioxide would occupy at the pressure of 1 atm (760 torr) and 25°C and the volume of the carbon dioxide dissolved in the acidic solution (V of acid × 0.80 mL) are additive (Gay-Lussac's Law of Combining Volumes).

3-3. From the total volume of CO_2 at 25°C and 760 torr, calculate the moles of CO_2 liberated by sample of Alka-Seltzer using the ideal gas law. Write a balanced equation for the reaction of hydrogen carbonate (HCO_3^-) with hydrochloric acid to produce CO_2. Calculate the grams of HCO_3^- in each piece of Alka-Seltzer and the percent by mass of HCO_3^-. Average the percent-by-mass data and calculate the standard deviation and relative standard deviation of your average. Complete the Report Form.

Name (Print)   Date (of Lab Meeting)   Instructor

Course/Section

14   EXPERIMENT 14: ALKA-SELTZER®: AN APPLICATION OF GAS LAWS

Prelab Exercises

1. Define the "standard state for gases." If gases are usually not at "standard state," why does the phrase exist?

2. Propose a chemical equation for the reaction of sodium bicarbonate and hydrochloric acid to produce carbon dioxide. What other products would be formed?

3. Why is it important to protect the Alka-Seltzer tablets from moisture or the air?

4. Why are you asked to saturate the distilled water in the leveling bulb with carbon dioxide before you begin the experimental determinations?

5. How will you test the system for leaks?

6.   What volume will be occupied by 0.015 mole of an ideal gas at 21°C and 760 torr?

7.   A sample of gaseous carbon dioxide was collected by the displacement of water at 22°C and 736 torr. The volume under these conditions was measured to be 215 mL. How many moles of carbon dioxide are in this sample?

8.   Answer any question given by your instructor.

Date _____   **Student's Signature** _____

Name (Print) _____   Date (of Lab Meeting) _____   Instructor _____

Course/Section _____   Partner's Name (If Applicable) _____

Report Form

DATA

Data is collected first in your notebook.

Laboratory temperature: _____ °C   or _____ K

Barometric pressure: _____ torr   or _____ atm

Vapor pressure of water at the temperature of the lab: _____ torr

Partial pressure of CO_2 in the buret: _____ torr

Mass of a whole tablet: _____ g

Measured or Calculated Quantity	Exp. 1	Exp. 2	Exp. 3	Exp. 4
Mass of tablet fragment				
Final buret reading (mL)				
Initial buret reading (mL)				
Volume of CO_2 collected (mL) at laboratory conditions				
Volume of 6 M HCl (mL)				

Date _____   **Instructor's Signature** _____

LAB REPORT

Attach your lab report that includes the PROBLEM STATEMENT (actual), YOUR MODIFICATIONS OF PROCEDURES, DATA/ANALYSIS, and CONCLUSION.

In the analysis section, include the following for each trial:

> volume of CO_2 collected, corrected to 25°C and 1 atm,
>
> volume of CO_2 dissolved in the HCl at 25°C and 1 atm,
>
> total volume of CO_2 produced from the reaction at 25°C,
>
> moles of CO_2 produced from the reaction,
>
> mass of HCO_3^- required to produce the CO_2, and
>
> percent of HCO_3^- by mass of the tablet.

Show a set of sample calculations for Exp. 1.

Be sure to average the percent-by-mass for all trials and calculate the standard deviation and relative standard deviation, showing calculations.

Include a particle view (drawing) of the reactants and the products for the reaction between sodium bicarbonate and hydrochloric acid.

POSTLAB QUESTIONS

1. In the steps that you followed for the calculations of this experiment, you made a conversion from moles of CO_2 to moles of HCO_3^-; write an equation that illustrates the chemical process upon which that mole ratio was based.

2. In this experiment, what was the relationship between $NaHCO_3$ and HCO_3^-?

3. What is the relationship between the mass of HCO_3^- and the total volume of CO_2 produced from the reaction?

4. What fraction of the total volume of CO_2 was contributed by the volume of CO_2 considered to be dissolved in the acid solution? What influence would a 17% change in the amount of CO_2 dissolved in the acid solution have on the calculated total volume of CO_2?

5. Using the product label, what other substances besides $NaHCO_3$ are in the Alka-Seltzer tablet? Could any of the other components make up a larger fraction of the tablet's content than the $NaHCO_3$ did?

6. Compare the chemical reaction in this experiment to the chemistry that takes place when $NaHCO_3$ is used to treat an acid spill.

7. A piece of an Alka-Seltzer tablet weighing 0.2331 g generated 28.55 mL of carbon dioxide at standard temperature and standard pressure after correcting for water vapor. Calculate the percent hydrogen carbonate in the sample.

8. Is the volume of a reactant gas related to the mass of a product in a reaction? Explain why or why not.

9. Are the volumes of reactant gases related to the volume of product gases in a reaction at STP where all substances are gaseous? Explain why or why not.

Date _____ **Student's Signature** _____

Exploring Intermolecular Forces

A Guided Inquiry Experiment

INTRODUCTION

Molecules are held together by intramolecular forces (covalent and ionic bonds), but the forces that attract one molecule to another molecule are called intermolecular forces. Intermolecular forces (IMF) include London dispersion forces (LDF), dipole-dipole forces, and hydrogen bonding (H-bonding). London dispersion forces are temporary attractive forces between two molecules when the electron distribution is NOT symmetrical causing a temporary dipole to occur. Dipole-dipole intermolecular forces are those that hold one molecule with permanent dipoles to another molecule with dipoles by the negative pole of one molecule aligning with the positive pole of another. Hydrogen bonding occurs between two molecules that have hydrogen directly bonded to a fluorine, oxygen, or a nitrogen.

OBJECTIVES

In this experiment, you will investigate the relationship between intermolecular forces and physical properties, such as state of matter, viscosity of liquids, boiling or melting points.

CONCEPTS

This experiment uses the concepts of Lewis dot structures bond polarity, and molecular polarity.

TECHNIQUES

Use of molecular models, care of thermometers, and safety around volatile liquids are some of the techniques encountered in the experiment.

ACTIVITIES

You will observe a number of substances that differ in intermolecular forces. You will use these observations to draw generalities about the relationships between physical properties and intermolecular forces.

PROCEDURES

Examining Molecules

1-1. Obtain a model kit to construct the following molecules using the Lewis structures completed for the prelab: water, acetone, hexane, and ethanol (or the list from your instructor). Use the models to determine molecular shape.

1-2. Draw the 3-D sketch of each model in the table in your Report Form.

1-3. For water, determine the polarity of the bonds, and then determine the molecular polarity. Repeat for each molecule. Record your determinations.

Evaporation

2-1. Obtain four pieces of filter paper (2 cm × 5 cm), a thermometer, four rubber bands, and a stopwatch or other timekeeping device.

2-2. Attach one piece of filter paper to the end of the thermometer at the bulb with a rubber band.

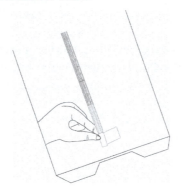

(a)   **(b)** Roll the paper around the thermometer.   **(c)** Secure with a rubber band.

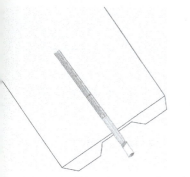

(d) Extend off benchtop to collect data.

2-3. Obtain a test tube containing water. Hold the thermometer in test tube, making sure the bulb is submerged in the liquid. Keep the thermometer in the liquid. When the temperature becomes constant, record this temperature as the time 0.0 minute.

2-4. Plan with your partner to begin timing every ½ minute immediately when you remove the thermometer from the liquid. Gently touch the side of the test tube if a drop clings to the filter paper to remove it. Also, plan to tape the thermometer with the attached filter paper to the top of the bench such that a 4 to 5 cm portion of the thermometer above the filter-covered bulb is extended off the bench. Be sure the thermometer is taped so that you can read the thermometer. Once your planning is done, go on to Procedure 2-5.

2-5. Remove the thermometer from the test tube and begin timing, tape it to the bench as planned. Record the temperature every 30 seconds for 8 minutes. You may stop recording early if the temperature stops decreasing and increases for two subsequent readings.

2-6. Repeat Procedures 2-3 through 2-5 for acetone, then for hexane, and then for ethanol. Your instructor may direct you to share test tubes with other groups or to test a different liquid with Procedures 2-3 through 2-5.

CAUTION

Acetone, hexane, and ethanol can be toxic by ingestion; avoid contact with your skin or prolonged inhalation. Wash your hands before leaving the laboratory. Alcohol and acetone are very flammable. There should be no open flames nearby.

Reaction to Charge	**3-1.**	There are four burets containing water, acetone, hexane, and ethanol. Below each buret is a 400-mL beaker. There is also a funnel in the top of each buret.
	3-2.	Create a static charge on a plastic rod (or pipe) by rubbing it with a cloth or fur.
	3-3.	Open the stopcock on each of the burets so that a slow steady stream is flowing into the beaker below.
	3-4.	Slowly bring the rod horizontally in front of each stream. Observe carefully for any deflection or bending of the streams. Repeat to determine which deflects the most. Rank each liquid. Record your results.
	3-5.	If needed to get the rankings of the liquids, repeat Procedure 3-4 with a recharged rod. (Recharge the rod by repeating Procedure 3-2.)
	3-6.	Close the stopcock of the buret to stop the flow of liquid. Cleanup by pouring the contents of the beakers into the burets through the funnel.
Viscosity	**4-1.**	Obtain a set of tubes containing various liquids, each with a marble (bead) inside.
	4-2.	Invert one of the tubes and observe the marble (bead) as it falls.
	4-3.	Plan with your partner to time the rate of fall of the marble (bead). Do this by inverting the tube and recording the number of seconds required for the marble (bead) to travel to the bottom of the tube.
	4-4.	Record both your observations and the time for the marble (bead) to reach the bottom of the tube.
Phases	**5-1.**	Obtain a set of Erlenmeyer flasks containing various hydrocarbons.
	5-2.	Swirl each flask. Record your observations of the phase. If it is a liquid, make an estimate of the viscosity.
	5-3.	Obtain a set of Florence flasks containing various substances with the same ratio of carbon to OH.
	5-4.	Swirl each flask. Record your observations of the phase. If it is a liquid, make an estimate of the viscosity.
Cleaning	**6-1.**	Put all materials away and clean your laboratory space, as your instructor directs.
	6-2.	Have your data sheets signed by your instructor.

15   EXPERIMENT 15: EXPLORING INTERMOLECULAR FORCES

Prelab Exercises

1. Define viscosity.

2. What is meant by a hydrocarbon?

3. Define evaporation.

4. List the possible safety hazards in this experiment. For each hazard, list what you should do to minimize these hazards.

5. In the Data Table for Examining Molecules, draw each Lewis structure. Your instructor will check this and sign below the table before you can begin the experiment.

6. Draw the Lewis dot structures for methane, butane, and pentane. Label each.

7. Answer any question assigned by your instructor.

Date _____**Student's Signature**_____

15   EXPERIMENT 15: EXPLORING INTERMOLECULAR FORCES

Report Form

DATA

Examining Molecules

	Lewis Structure	Geometry or Shape of the Molecule	Polar Bonds? Yes or No	Polar or Nonpolar Molecule?
Water H_2O structural formula HOH				
Acetone C_3H_6O structural formula CH_3OCH_3				
Hexane C_6H_{14} structural formula $CH_3(CH_2)_4CH_3$				
Ethanol C_2H_6O structural formula CH_3CH_2OH				

Date _____ **Instructor's Signature** _____

(Lewis structures are done before the experiment begins)

Evaporation

	Water T (°C)	Acetone T (°C)	Hexane T (°C)	Ethanol T (°C)
0.0 min				
0.5 min				
1.0 min				
1.5 min				
2.0 min				
2.5 min				
3.0 min				
3.5 min				
4.0 min				
4.5 min				
5.0 min				
5.5 min				
6.0 min				
6.5 min				
7.0 min				
7.5 min				
8.0 min				

Reaction to Charge

	Observation	Ranking: 1 = most deflected, 4 = least deflected
Water		
Acetone		
Hexane		
Ethanol		

Viscosity

	Observation	Ranking: 1 = fastest, 4 = slowest
Water		
Oil		
Hexane		
Ethanol		

Phases-Hydrocarbons

Substance	MP (°C)	BP (°C)	State of Matter	# of Carbon	Viscosity
Methane CH_4	−183	−164			
Butane $CH_3(CH_2)_2CH_3$	−190	−0.5			
Pentane $CH_3(CH_2)_3CH_3$	−130	36			
Hexane $CH_3(CH_2)_4CH_3$	−95	69			
Mineral Oil (paraffin oil) $CH_3(CH_2)_{16}CH_3$	−	260–330			
Paraffin Wax $CH_3(CH_2)_{22}CH_3$	47–65	>400			

Phases-C to OH Ratio

Substance	MP (°C)	BP (°C)	State of Matter	# of Carbon	# of OH	Viscosity
Methanol	−98	65				
Ethylene Glycol	−12.9	197				
Glycerol	17.8	290				
Glucose	149–152	–				

Date _____ **Instructor's Signature** _____

ANALYSIS
Examining Molecules
1. How are polar bonds related to polar molecules?

2. Which, if any, molecules are nonpolar?

3. Which, if any, molecules are polar?

4. Give the strongest type of intermolecular forces present in each liquid. Explain your reasoning.
    Water

    Acetone

    Hexane

    Ethanol

Evaporation
1. What was the change in temperature for each of the liquids (use the stable starting temperature and the lowest temperature while timing)?

Water ΔT (°C)	Acetone ΔT (°C)	Hexane ΔT (°C)	Ethanol ΔT (°C)

2. What is the strongest intermolecular force of the liquid that had the largest temperature change?

3. What is the strongest intermolecular force of the liquid that had the smallest temperature change?

4. Temperature change is a measure of the rate of evaporation. Which liquid has the largest change in temperature?

5. What relationships do you see between the strength of intermolecular force and temperature change?

6. What is the relationship between the strength of intermolecular forces and evaporation rate?

Reaction to Charge

1. What is the strongest intermolecular force of the liquid that had the most deflection?

2. What is the strongest intermolecular force of the liquid that had the least deflection?

3. What is the relationship between the strength of intermolecular forces and amount of deflection?

Viscosity

1. Give the strongest type of intermolecular forces present in mineral oil. Explain your reasoning. (You have already done this for the other substances.)

2. Which substance is the most viscous?

3. What is the relationship between the strength of intermolecular forces and viscosity?

Phases

1. Give the strongest type of intermolecular forces present in each substance. Explain your reasoning. (You have already done this for hexane and mineral oil.)
    Methane

    Butane

    Pentane

    Paraffin wax

2. Which hydrocarbon has the stronger intermolecular forces? Explain your choice.

3. Do you see any pattern between the structure and the intermolecular force for hydrocarbons? Explain the pattern.

4. What pattern(s) do you see with boiling points for the hydrocarbons?

5. Give the strongest type of intermolecular forces present in each substance. Explain your reasoning.
   Methanol

   Ethylene Glycol

   Glycerol

   Glucose

6. What is the carbon to OH ratio in the four substances you tested? Explain your reasoning.

7. Which of the four substances has the stronger intermolecular forces? Explain your choice.

8. Do you see any pattern between the structure and the intermolecular force for these four substances with carbon and OH? Explain the pattern.

9. What pattern do you see with boiling points for the substances with carbon and OH?

10. Which has stronger intermolecular forces, a hydrocarbon or a substance with carbon and an OH group? Explain your choice.

11. What pattern do you see between intermolecular forces and boiling points?

12. What pattern do you see between intermolecular forces and phase?

POSTLAB QUESTIONS

1. What types of intermolecular forces did you find in the molecules you constructed from models?

2. Which should evaporate more quickly, butanol ($CH_3CH_2CH_2CH_2OH$) or pentane ($CH_3CH_2CH_2CH_2CH_3$)? Explain your choice.

3. Which would deflect the most when a charged rod is brought close, hexane ($CH_3CH_2CH_2CH_2CH_2CH_3$) or pentanol ($CH_3CH_2CH_2CH_2CH_2OH$)? Explain your choice.

4. Which would you predict to be more viscous, pentane ($CH_3CH_2CH_2CH_2CH_3$) or butanone ($CH_3CH_2COCH_3$)? Explain your choice.

5. Which would you predict to be a viscous liquid and which would you predict to be a solid, butanol ($CH_3CH_2CH_2CH_2OH$), 1,4-butanediol ($CH_2(OH)CH_2CH_2CH_2OH$), or 1,2,4-butantriol ($CH_2(OH)CH(OH)CH_2CH_2OH$)? Explain your choices.

6. Which would you predict to boil at a higher temperature, C_2H_6, CH_3CH_2Cl, or CH_3OH? Explain your choices.

7. Which would you predict to have stronger intermolecular forces, C_3H_8, $CH_3CH_2C(=O)H$, or CH_3CH_2OH? Explain your choice.

8. Rank the intermolecular forces by strength from strongest to weakest and give an example from this experiment that demonstrates this ranking.

9. Answer any questions assigned by your instructor.

Date _____ **Student's Signature** _____

Freezing Points of Solutions

A Guided Inquiry Experiment

INTRODUCTION

To de-ice roads or to force ice to melt in an ice cream maker, we sprinkle salt on the ice. Freezing point is a physical property of a compound or mixture. If a substance changes from a liquid to a solid, we call that freezing; but when the solid becomes a liquid, it is referred to as melting. Actually both the freezing and the melting occur over a range of temperatures. The melting point range is from the temperature that the first crystal liquefies to the temperature that the last crystal melts. The freezing point range is from the formation of the first crystal to the solidification of the last of the liquid. The freezing point range and the melting point range are usually small and equal for homogeneous systems. However, some systems tend to be slow in starting to crystallize. That is: they tend to form super cooled liquids that will not crystallize until a seed crystal is added. Because of super cooling, melting points are usually measured and sometimes referred to as freezing points. The freezing points that you are to find in this experiment are actually the midpoints of the melting point ranges.

OBJECTIVES

During this experiment, you will prepare a melting point determination setup like that described in Procedure 3-2. You will prepare melting point determination tubes of pure vanillin, a mixture of vanillin and a known compound, and a mixture of vanillin and an unknown compound. The mixtures of known and of unknown compounds with vanillin will be converted into solutions by melting them and allowing the components to mix at the molecular level. The resulting solution will be cooled, allowed to crystallize, and the melting point of the resulting solution taken. Relationships among these values will be explored.

CONCEPTS

Preparation of solutions, solvent versus solute, determination of masses, and the determination of melting or freezing points are several of the concepts used during this experiment.

TECHNIQUES

You will use thermometers to determine melting points. You will also make extensive use of the analytical balance. The Bunsen burner will be used to heat a water-filled melting point determination bath.

ACTIVITIES

In this experiment, you will determine the relationship between the amount of solute and the freezing point. You will be given an unknown, and you are to determine its molar mass using freezing point data that you collect.

CAUTION	

Under no circumstances are any of the chemicals used in this experiment to be tasted, swallowed, or taken from the laboratory. Many of these chemicals are toxic and all are hazardous when used without supervision. Be careful also in working with the hot equipment. In case of contact with hot glass or water, soak the affected area immediately with cold water and notify your instructor. Wear approved eye protection at all times.

PROCEDURES

Preparing the Melting Point Tubes

1-1.

NOTE: If the micro test tubes are provided to be used as the containers in which the melting points of the solutions of vanillin and known or unknown will be determined, proceed to Procedure 2-1.

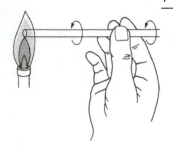

If you are asked to prepare the needed tubes, use the hottest flame you can obtain with your burner to seal one end of a 10- to 30-cm length of 6-mm (inner diameter) soft-glass tubing or a Pasteur pipet, which you have cut off at a point near where the small diameter portion begins to expand. Slowly rotate the tube while holding it just above the blue inner flame. The hottest part of the flame is just above the inner blue cone. Heat the tube until the end is sealed.

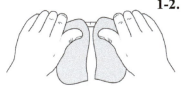

1-2. To cool the hot tube, lean it against a ringstand base so that the hot end is above the base. *Do not place hot glass on the top of the bench or in water.* When the tube has cooled, place it flat on the bench. Use a file to scratch the tube about 4 cm from the sealed end. Carefully snap the tube at the scratch mark by placing your thumbs on the side behind the scratch and pushing away with your thumbs.

CAUTION	

You may wish to protect your hands with a towel.

1-3. Repeat Procedures 1-1 and 1-2. You will need three melting point tubes.

Filling the Melting Point Tubes

2-1. Take a piece of smooth paper (approximately 4 × 6 cm) to the analytical balance. Place approximately 110 mg vanillin (acceptable range: 100 to 120 mg) onto the paper. Be sure to record the exact mass of vanillin.

or $C_8H_8O_3$

2-2. Your vanillin sample should consist of small particles. If larger particles are present, divide them into smaller pieces with your spatula or crush with the end of a clean test tube. Large pieces may cause the experiment to fail.

2-3. Use your spatula to transfer the sample of vanillin into the melting point tube. Then gently tap the closed end of the tube against the benchtop to help the crystals slide to the bottom of the tube. You should continue to fill and compact until as much of the sample as possible has been transferred. Place the filled tube upright in a small beaker or holder.

2-4. Take two pieces of smooth paper (approximately 4 × 6 cm), the stock bottle of vanillin, and a stock bottle of lauric acid to the analytical balance. Again tare the balance and one piece of paper. Place approximately 110 mg vanillin (acceptable range: 100 to 120 mg) onto this paper.

Tare another piece of smooth paper (approximately 4 × 6 cm). Add an amount of lauric acid corresponding to approximately 5% (acceptable range: 5 to 12%) of the mass of vanillin. Record the masses of the vanillin and of the lauric acid.

2-5. Transfer the lauric acid sample as quantitatively as possible onto the paper containing the vanillin. Thoroughly mix the combined sample of vanillin and lauric acid. All larger particles must be crushed into small particles. Use your spatula and the end of a clean test tube to crush and mix your sample. Insufficient mixing may yield misleading results.

NOTE: Thorough mixing is vital to the experiment.

2-6. Use your spatula to transfer the vanillin and lauric acid mixture into a clean melting point or micro-test tube. You can gently tap the tube with your finger to assist the material in sliding into the tube. As much of the mixture as possible needs to be transferred in the micro test tube. Mark the tube so that you can distinguish it from the others. You can do this by attaching tape to the mouth of the tube. Store this tube in the beaker or holder with the first tube containing pure vanillin.

2-7. Obtain a container of an unknown from your instructor. Take two pieces of smooth paper (approximately 4 × 6 cm), your unknown sample, and the stock bottle of vanillin to the analytical balance. Again tare the balance and paper. Place approximately 110 mg vanillin (acceptable range: 100 to 120 mg) onto one paper.

2-8. Tare another piece of smooth paper. Add an amount of unknown corresponding to approximately 5% (acceptable range: 5 to 12%) of the mass of vanillin. Record the mass of the vanillin and the unknown.

2-9. Transfer the unknown sample onto the paper containing the vanillin. Mix the combined sample thoroughly and crush all larger particles into smaller pieces. Use your spatula and the end of a clean test tube to crush and mix your sample. Again, insufficient mixing may cause the experiment to fail.

2-10. Use your spatula to transfer the vanillin and unknown mixture into the third micro test tube. Mark the tube so that you can distinguish it from the one filled in Procedure 2-3 that contains pure vanillin and from the one filled in Porcedure 2-6 that contains the mixture of vanillin and lauric acid. Labeled tape near the mouth of the tube will work. Store this third tube in the beaker or holder.

Melting Points

3-1. Slide a large cork with a center hole and a pie-sized section cut away onto a 110-degree thermometer (or a thermocouple probe).

3-2. Attach an iron ring to a ringstand at a height that allows space for a burner to be placed under it. Place a wire gauze on the ring. A 500-mL Erlenmeyer flask with about 400 mL of water can serve as your melting point bath. Attach a clamp securely—but not too tightly—to the neck of the 500-mL Erlenmeyer flask. Place the flask on the wire gauze on the ring clamp and attach the clamp holding the flask to the ringstand. Check that the flask is securely fastened to the stand and cannot slip off the ring.

3-3. Use a rubber band to secure the three melting-point tubes to the thermometer (or a thermocouple probe). The rubber band must be near the top of the shortest tube and the bottoms of the tubes should be near the tip of the thermometer (or thermocouple). Position the melting-point tubes as shown in the illustration.

3-4. Clamp the split cork and thermometer into position. Slide the thermometer clamp down the ringstand until the thermometer bulb (or thermocouple probe) is immersed in the water. Tighten the clamp to the ringstand. Begin to heat the water.

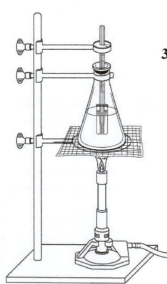

Practice changing the air and gas controls of the burner to change the rate of heating. Also vary the location of the burner to find the best position for causing convection currents in the water.

Raise the thermometer (or thermocouple probe) and tubes as the water level rises. You may heat rather rapidly to a temperature around 70°C, but above 65°C the temperature must not rise faster than 1°C per 30 seconds. Watch only the PURE vanillin tube at this stage.

Record the temperature at which liquid first appears in the tube and the temperature at which the last crystals of the vanillin melts. The average of these two temperatures will be the melting point value that you will use in your calculations. *If your heating rate was sufficiently slow and the vanillin is pure, the difference between the temperature at which the first liquid appeared and the temperature at which the last crystal melted will be only 1 to 2 degrees.*

3-5. After you have determined the melting point of pure vanillin, heat the water to 90° but not higher than 95°. Hold this temperature until the vanillin/lauric acid and vanillin/unknown mixtures have formed clear, homogeneous liquids. Lift the thermometer (or thermocouple probe) and tubes out of the water. Clamp them so that they are just above the surface of the hot water. (If your ring stand is not tall enough to allow you to position the thermometer bulb and the attached tubes "just above the surface of the hot water" rotate the clamp so that they are positioned outside of the flask.) Let the water slowly cool to approximately 60°.

NOTE: If you heated your sample to a temperature higher than 95°C or allowed it to remain in the hot water for an extended length of time, some vanillin may have condensed to the cooler part of the tube. Check the top part of the micro-test tubes for vanillin crystals. If a considerable amount of vanillin has sublimed, you may have to prepare another vanillin/lauric acid or vanillin/unknown mixture. Consult your instructor.

3-6. The vanillin mixtures should solidify by the time the water temperature has dropped to 55°C. (If not, consult your instructor.) When all tubes contain solids, lower the thermometer (or thermocouple probe) and tubes back into the water. Begin reheating the water at a rate of about 5°C per minute. Very closely watch the vanillin mixtures in the micro test tubes. Record the temperature at which liquid first appears and the temperature at which the last cloudiness due to solid crystals disappears.

These two temperatures provide you with the *approximate* melting range of each mixture. **Record this as your first trial and your approximate melting range.** Turn off the burner when the last solid has melted. Again, raise the thermometer and tubes to just above the surface of the water. Allow the water to cool until the solutions have solidified.

3-7. When the solution in the bulb has solidified, lower the thermometer and tubes and reheat the water at a rate of 5 to 10°C per minute to a temperature 10°C below your "approximate" melting range determined in Procedure 3-6. Then reduce the heating rate to 1°C per minute and determine the actual melting point range for both the vanillin/lauric acid and vanillin/unknown mixtures. **Record this as Trial 2.**

3-8. If the melting points of the solutions are significantly lower for Trial 2, it may be the result of incomplete mixing the first time they were melted. In this case, turn off the burner and raise the thermometer and tubes to just above the surface of the water. Allow the water to cool until the solutions have solidified. Then repeat Procedure 3-7. **Record this as Trial 3 and disregard Trial 1 when you do the calculations.** In the rare event that Trials 1 and 2 are very close, you do not have to do a third trial.

3-9. If time permits, complete another determination of the vanillin/unknown mixture using fresh samples and new glass tubes. This determination is referred to as "New Determination" on the Report Form. If you are going to make another determination with a new mixture, you can prepare this mixture while the water is cooling.

3-10. When you have finished your work, allow the water to slowly cool while you clean up your work area. Dispose of the unused mixture and the melting tubes by placing them in the designated containers. Have your instructor sign your notebook and Report Form.

Calculations

4-1. Use the mid-point of the melting point ranges as the melting points of the pure vanillin, the vanillin/lauric acid, and vanillin/unknown samples.

16   EXPERIMENT 16: FREEZING POINTS OF SOLUTIONS

Prelab Exercises

1. What is the major distinction between melting point and freezing point?

2. Calculate the molar mass of lauric acid ($C_{12}H_{24}O_2$). Record this for use in the lab report.

3. In which area of the flame of a Bunsen burner is the temperature lowest?

4. In the note after Procedure 3-5, directions are given to be followed if a considerable amount of vanillin has sublimed. What has happened when vanillin is sublimed?

5. Answer any questions assigned by your instructor.

Date _____ **Student's Signature** _____

Name (Print) ______ Date (of Lab Meeting) ______ Instructor ______

Course/Section ______ Partner's Name (If Applicable) ______

16 EXPERIMENT 16: FREEZING POINTS OF SOLUTIONS

Report Form

DATA

Data is collected first in your notebook.

Mass of vanillin: ______ mg

Melting range of vanillin: ______ °C

Data from other groups:

Table 1 *Pure Vanillin*

Mass of Vanillin	Melting Pt. Range	Melting Point	Mass of Vanillin	Melting Pt. Range	Melting Point

Composition of the vanillin/lauric acid solution:

Mass of vanillin: ______ mg

Mass of lauric acid: ______ mg

Melting point range of vanillin/lauric acid solution: trial 1 ______ °C

trial 2 ______ °C

trial 3 ______ °C

Data from other groups:

Table 2 *Vanillin/Lauric Acid Mixture*

Mass of Vanillin	Mass of Lauric Acid	Melting Pt. Range	Melting Point	Mass of Vanillin	Mass of Lauric Acid	Melting Pt. Range	Melting Point

Composition of the vanillin/unknown solution:

Mass of vanillin: _____ mg

Mass of unknown: _____ mg

Unknown number or code name: _____

Melting point range of vanillin/unknown acid solution: trial 1 _____ °C

trial 2 _____ °C

trial 3 _____ °C

New determination, if time permits:

Mass of vanillin: _____ mg

Mass of unknown: _____ mg

Unknown number or code name: _____

Melting point range of vanillin/unknown acid solution: trial 1 _____ °C

trial 2 _____ °C

Date _____   **Instructor's Signature** _____

ANALYSIS

1. Determine the melting point from your individual data for PURE vanillin: _____ °C

2. How does your melting point for PURE vanillin compare with those found by other groups?

3. The melting point of PURE vanillin, based on all the data you have (yours and other groups): _____ °C. Explain how you determined this number.

4. Does the melting point of PURE vanillin depend on the mass of vanillin used? Explain your answer using the data you collected.

5. Determine the melting point from your individual data for vanillin/lauric acid mixture: _____ °C.

6. How does your melting point for vanillin/lauric acid compare with those found by other groups?

7. How do the melting point for PURE vanillin and that of the vanillin/lauric acid mixture compare?

8. If you consider the vanillin/lauric acid mixture to be a solution, what is the solute and what is the solvent?

9. The traditional concentration unit used in this situation is molality (m). Molality is **moles of** solute per kg of solvent. Find the molality for the vanillin/lauric acid mixtures. Find the difference in melting point of each mixture in Table 2 from the average melting point from that for PURE vanillin ($\Delta T = T_{mixture} - T_{pure}$).

Table 3 *Molality and Change in Temperature of Each Mixture in Table 2*

Molality (m)	Change in Melting Pt.	Molality (m)	Change in Melting Pt.	Molality (m)	Change in Melting Pt.

Show your calculations:

10. What relationship is there between the molality of the vanillin/lauric acid mixture and the change in the melting point from PURE vanillin? Do you find any general trends or constants when you try different mathematical operations with the numbers ($+, -, \times, /$)? Give a table of your work. Graph the data. If practical, set the y-intercept at zero. Attach the graph and give the slope if you find a linear relationship. What is the average relationship? Give the relationship you find, explain your thinking, and show calculations.

11. Use your textbook to find the name for the relationship/constant/slope you found. What are the units on this value?

12. Assuming that the same relationship exists between the molality and the change in melting point, what is the molality of your vanillin/unknown mixture? (Show calculations.)

13. Approximate molar mass of the unknown: _____
(Show calculations.)

14. If your unknown were known to be either benzoic acid ($C_7H_6O_2$) or palmitic acid ($C_{16}H_{32}O_2$), which one would your data suggest? (Show calculations.) What is the percent error if that is actually your unknown?

POSTLAB QUESTIONS

1. Give the common units for:

a. K_f

b. m

c. ΔT_f

2. The melting point of pure camphor is 177°C; its constant is 37.7°C kg/*m*. Calculate the melting point of a solution prepared from 120 mg of camphor and 0.000155 moles of a solute.

3. Explain how freezing points can be used to determine the molar mass of an unknown solute.

4. Draw a particle picture of a solution solidifying.

Date _____ **Student's Signature** _____

Spectrochemical Analysis

A Guided Inquiry Experiment

INTRODUCTION

Every day millions of spectrochemical analyses are carried out. Experiments that depend upon the measurement of mass and volume (gravimetric and volumetric determinations) are common in an introductory chemistry courses but, in this experiment, you will instead utilize the measurement of electromagnet radiation and its interaction with matter. The terms spectrochemical analysis or spectroscopy are often used interchangeably to describe this type of determination. Spectrochemical analysis is frequently the method of choice for both qualitative and quantitative determinations. These analyses range from the simple visual color tests, to tests of the absorbance of light of a specific wavelength, to x-ray analyses, and to 3-dimensional images such as CAT scans. Spectrochemical analysis has clinical, environmental, quality control, law enforcement, color matching, and novelty applications. All spectrochemical methods involve electromagnetic radiation and its interaction with matter.

It is customary to consider how a sample would interact with different wavelengths in a given region of electromagnetic radiation, and this collection of measurement signals as a function of wavelength is called a spectrum. The device used to conduct these analyses is called a spectrophotometer.

OBJECTIVES

In this experiment, the hydrated Cu^{2+} ion, $[Cu(H_2O)n]^{2+}$ is reacted with ethylenediamine to produce the complex ion, $[Cu(H_2NCH_2CH_2NH_2)_2]^{2+}$. The hydrated Cu^{2+} ion produces a very light blue solution that is not a strong absorber of light of any wavelength. However, the complex ion is deeply colored and thus a strong absorber. The wavelengths of light that are absorbed the most by solutions of this complex ion will be determined via an absorption curve to find the wavelength of maximum absorption. The relationship between absorption and concentration at this maximum wavelength will be explored.

CONCEPTS

This experiment uses the concepts of concentration, spectroscopy, transmittance, and absorbance. You may wish to review the Spectroscopy section in the Common Procedures and Concepts at the end of this manual.

TECHNIQUES

You will use your lab skills to follow directions, prepare solutions, make dilution calculations, operate an electronic instrument, and to organize your work area and data records. Gravimetric, volumetric, and spectrophotometric methods all utilize good basic laboratory techniques.

ACTIVITIES

You will prepare standard solutions of ethylenediamine-Cu^{2+} complex, determine the absorption curve of the Cu-ethylenediamine complex, the wavelength of the maximum, and any relationship between absorption and concentration. You are asked to analyze two unknown solutions of ethylenediamine-Cu^{2+} complex.

> **CAUTION**
>
> **You will be working with some potentially dangerous reagents. Wear approved eye protection at all times. In case of contact with chemicals, wash the affected areas immediately with plenty of tap water.**

PROCEDURES

Preparation of Solutions

1-1. In order to analyze your unknowns, you will need to investigate the interaction of known concentrations with a spectrophotometer. First you will need to prepare these solutions. Set up three clean burets and label them "Cu^{2+}," "H_2O," and "en" (ethylenediamine).

1-2. Calculate the amount of $CuSO_4 \cdot 5H_2O$ that is needed to prepare 100 mL of a 0.010 M solution of $CuSO_4$. Tare the analytical balance with a piece of smooth paper on the pan. Transfer to the paper an amount of $CuSO_4 \cdot 5H_2O$ that is between +15 mg and +25 mg of the amount calculated above. Record the mass. Leave the balance and its surroundings as clean or cleaner than you found them.

1-3. Place a funnel into a clean but not necessarily dry 100-mL volumetric flask. Transfer all the copper sulfate from the smooth weighing paper into the flask. Rinse the last traces of copper sulfate from the funnel into the flask with a stream of distilled water from your wash bottle. Fill the flask about half full with distilled water. Swirl the flask until all the $CuSO_4 \cdot 5H_2O$ has dissolved.

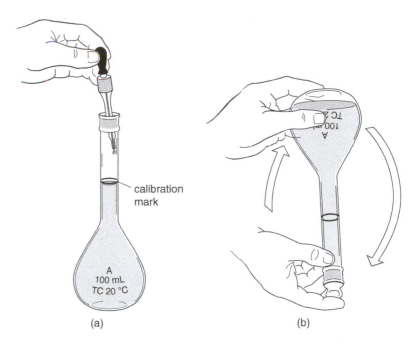

(a) (b)

Then fill the flask with distilled water to the 100-mL mark. Snugly stopper the flask. While holding in the stopper, invert the flask 10 to 20 times to produce a homogeneous solution. Calculate the concentration of the Cu^{2+} solution actually prepared. (REMEMBER: You didn't add the amount of solute calculated to produce exactly 0.010 M solution.)

1-4. Rinse the "Cu^{2+}" buret twice with 5-mL portions of the Cu^{2+} solution. Fill this buret to the 30-mL mark with the Cu^{2+} solution.

Rinse and fill the second buret with distilled water.

Rinse the third buret with a small volume of the 1.0 M ethylenediamine (en) solution provided by your instructor. Place no more than 5 mL of en in the "en" buret.

Drain enough liquid from each buret into a "waste" flask to expel air from the buret tips.

1-5. Clean and dry seven test tubes (8 mL or larger). Label the test tubes "1," "2," . . . "7." Drain from the "Cu^{2+}" buret into test tubes "2" through "7" the volume of Cu^{2+} solution given in the table. Record each volume to the nearest 0.01 mL.

1-6. To each of the test tubes "1" through "7," add 0.5 mL of the 1.0 M solution of ethylenediamine. Then add the amount of distilled water needed to bring the total volume of liquid in each test tube to 7.00 mL. Record all volumes to the nearest 0.01 mL.

Test Tube	mL Cu^{2+}
2	0.5
3	1.0
4	2.0
5	3.0
6	4.0
7	5.0

Volume (mL)

Tube #	Cu²⁺	en	H₂O	Total
1	—	0.5	6.5	7.0
2	0.5	0.5	6.0	7.0
3	1.0	0.5	__.__	7.0
4	2.0	0.5	__.__	7.0
5	3.0	0.5	__.__	7.0
6	4.0	0.5	__.__	7.0
7	5.0	0.5	__.__	7.0

1-7. Tightly stopper each test tube with a clean and dry stopper. Invert each tube several times to thoroughly mix. Set the test tubes into a test tube rack. Calculate the concentrations of copper in each of the solutions in each test tube before any reaction takes place.

Absorption Curve of the $[Cu(en)_2]^{2+}$ Complex

2-1. Beginning at 400 nm, set the zero-transmittance and the 100% transmittance (as described in the Spectroscopy section of Common Procedures and Concepts Section at the end of this manual) with an empty chamber and a water-filled cuvet. Next place a solution from test tube "7" in the chamber. Record the percent transmittance and absorbance. Record both values in your notebook.

2-2. Move the wavelength selector knob to 420 nm. Repeat Procedure 2-1, except this time the wavelength is set to 420 nm. Read the percent transmittance and absorbance.

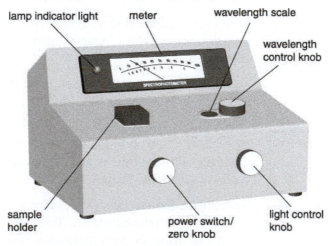

2-3. Repeat Procedure 2-1, but this time advance the wavelength selector knob another 20 nm. Continue likewise with 20 nm advances until you have reached 600 nm. You may keep solution 7 in the cuvet for use in Procedure 3-1.

2-4. Plot absorbance (y-axis) versus wavelength (x-axis). Locate the maximum of the absorption curve of $Cu(en)_2^{2+}$. Assume that all the Cu^{2+} was present as the en-complex.

Calibration Curve

3-1. Set the wavelength selector at the wavelength corresponding to the maximum of the absorption curve. Using the cuvet containing only distilled water, adjust "0" and "100%" transmittance (see Common Procedures and Concepts). Place the $Cu(en)_2^{2+}$ solution (test tube "7") back into the beam and measure the transmittance and absorbance. Record the test tube number and the transmittance and absorbance.

3-2. Remove the cuvet with the $Cu(en)_2^{2+}$ solution from the sample chamber. Discard the contents in the designated waste container.

Place a small volume of solution from test tube "6" into the cuvet. Swirl the solution in the cuvet. Then discard the solution from the cuvet as directed by your instructor. Rinse the cuvet again by placing another small volume of solution from test tube "6" into the cuvet. Swirl the solution in the cuvet. Then again discard the solution from the cuvet as directed by your instructor. Fill the cuvet with solution from test tube "6."

Place the cuvet with the $Cu(en)_2^{2+}$ solution into the light beam and read and record the transmittance and absorbance. *Absorbance is the value that is usually preferred during subsequent calculations.* (One may obtain better results by measuring the transmittance and calculating the absorbance from the transmittance.)

3-3. Repeat Procedures 3-1 and 3-2 with the solutions from test tubes "5," "4," "3," "2," and "1."

3-4. Construct a calibration curve by plotting absorbance versus copper concentrations in the solutions.

Determination of Unknown Copper Concentrations

4-1. Now, compare the interaction with the spectrophotometer of the unknowns with those of the known solutions. Take two clean, dry test tubes to your instructor. From the assigned burets, place 2.0 mL of Cu^{2+} unknown in each test tube. The test tubes will be labeled with a code "U-#" and a volume.

4-2. To each of your two unknowns add 0.5 mL of the 1.0 M ethylenediamine solution and 4.5 mL of distilled water to bring the total volume of solution to 7.0 mL. Stopper the test tubes and invert them several times.

4-3. Repeat Procedures 3-1 and 3-2 but this time use the unknowns to measure the transmittance and absorbance of these solutions at the wavelength corresponding to the maximum in the absorption curve (Procedure 2-4).

4-4. Clean the cuvets, burets, and all other glassware. Switch the spectrophotometer off, etc., as directed by your instructor.

4-5. Have your instructor sign your notebook and Report Form. Complete the Report Form.

Detection Limits (optional)   **5-1.** On the Report Form you are asked to determine the "Detection Limits" in this experiment. At this time, you will need to look closely at the spectrophotometer and record in your notebook the smallest absorbance value that you can logically measure with this instrument. (Discuss with your instructor and other students to determine if they have a different interpretation of "smallest absorbance value that you can measure.")

Name (Print)   *Date (of Lab Meeting)*   *Instructor*

Course/Section

17   EXPERIMENT 17: SPECTROCHEMICAL ANALYSIS

Prelab Exercises

1. Give the definitions for each (you may wish to check the section on spectroscopy in the Common Procedures and Concepts Section).

   a. molarity

   b. percent transmittance

   c. absorbance

2. Give the equation for a straight line.

3. Calculate the mass of $CuSO_4 \cdot 5H_2O$ required for preparation of 100. mL of 0.010 M Cu^{2+} solution. (Record your work and answer for use in Procedure 1-2.)

4. Describe three techniques used when using the spectrophotometer. (See Common Procedures and Concepts Section at the end of this manual.)

5. What is a cuvet?

Date _____   **Student's Signature** _____

Name (Print)   Date (of Lab Meeting)   Instructor

Course/Section   Partner's Name (If Applicable)

17   **EXPERIMENT 17: SPECTROCHEMICAL ANALYSIS**

Report Form

DATA

Mass of $CuSO_4 \cdot 5H_2O$ used to prepare standard Cu^{2+} solution: _____ g

Concentration of standard Cu^{2+} solution prepared: _____ *M*

Wavelength versus percent transmittance and absorbance (Procedures 2-1 through 2-4):

nm	400	420	440	460	480	500	520	540	560	580	600
%T	___	___	___	___	___	___	___	___	___	___	___
A	___	___	___	___	___	___	___	___	___	___	___

Wavelength of maximum absorption used for Calibration Curves (Procedure 2–4): _____

Calibration Curve and Unknowns:

Tube No.	1	2	3	4	5	6	7	U# ___	U# ___
Vol. Cu^{2+} Soln. (mL)									
Vol. "en" Reagent (mL)									
Vol. Distilled Water (mL)									
New Cu^{2+} Conc. Units = ( )									
Transmittance Units = ( )									
Absorbance Units = ( )									

Date _____   **Instructor's Signature** _____

ANALYSIS

1. Attach a sheet of graph paper on which you have plotted your data for Procedures 2-1 through 2-4. Absorbance should be the vertical or y-axis and the wavelength the horizontal or x-axis. Locate the maximum of the absorption curve of $Cu(en)_2^{2+}$. Assume that all the Cu^{2+} was present as the en-complex.

    Wavelength that gave maximum absorbance = _____ nm

2. Find the relationship between the concentrations and the absorbances. Attach a sheet of graph paper on which you have plotted your data for absorbance versus copper ion concentration (Procedures 3-1 through 3-4). This is known as a **CALIBRATION CURVE**. Find an algebraic equation to express the relationship between these two variables. What would the units be on any constant in your equation?

3. Using the algebraic equation from the previous question, what are the concentrations of your two unknowns?

    U# _____ = _____ *M*

    U# _____ = _____ *M*

4. What were the concentrations of your two unknowns before dilution to 7 mL?

    U# _____ = _____ *M*

    U# _____ = _____ *M*

5. Draw particle pictures of the maximum wavelength light passing through tube "7" vs. tube "2".

6. Detection Limits (optional)

Smallest absorbance measurable with the spectrophotometer in your laboratory (Procedure 5-1):

$A_{smallest}$ _____

Use the data from the absorption curve of the $[Cu(H_2NCH_2CH_2NH)_2]^{2+}$ complex to calculate the detection limit for the determination of copper using the color of the Cu^{2+}-ethylenediamine complex. (Show your calculations.)

Detection limit _____ mol L^{-1}

_____ g Cu^{2+} per liter

POSTLAB QUESTIONS

1. Describe the relationship between absorbance and concentration for a clear but colored solution. Would you expect a similar relationship between concentration and transmittance? Explain.

2. Absorbance depends on the thickness of the sample size. The cuvets for most spectrophotometers hold the sample size to 1 cm, although they can vary. The molar absorptivity (ε) is equal to the absorbance of a 1-cm-thick sample of a 1.0 M solution. In this case, ε has the units L mol^{-1} cm^{-1}. ε changes with wavelength and is generally given for the wavelengths corresponding to the maxima in the absorption curve. How is the constant you found related to the molar absorptivity?

3. A solution known to contain 0.024 M Cu^{2+} forms a complex with ethylenediamine that yields an absorbance of 0.37 in a cell of 1.0 cm path length at a selected wavelength. Calculate ε for these conditions.

4. Calculate the concentration of an unknown copper solution with an absorbance of 0.49 under the same conditions as given in Postlap Question 3.

5. From the absorption curve of $Cu(en)_2^{2+}$ you constructed in the first analysis question, fill in the following. (Show your work or give your reasoning.)

the wavelength of maximum absorption: _____ nm

the molar absorptivity ε at the maximum: _____ $L \; mol^{-1} \; cm^{-1}$

the molar absorptivity 50 nm past the
    maximum (toward longer wavelength): _____ $L \; mol^{-1} \; cm^{-1}$

the molar absorptivity 50 nm before the
    maximum (toward shorter wavelength): _____ $L \; mol^{-1} \; cm^{-1}$

Show your work.

6. Why is it important that the line on the cuvet be aligned with the guide on the spectrophotometer?

7. Why must the zero absorbance and 100% transmittance be readjusted each time the wavelength is changed?

8. What role does the solution in tube "1" play in this experiment?

Date _____ **Student's Signature** _____

Heat of Crystallization

A Skill Building Experiment

INTRODUCTION

Energy, its cost, availability, and conservation are frequent topics of conversation. In chemistry, you have been taught that energy is never used up. Energy is, however, converted from one form to another. The later form of energy is often less available for useful work. In this experiment, you will study a system that has the potential to capture solar energy and convert the energy into a stored form that can be released as thermal energy when needed to heat a home or office.

Solar energy is an inexpensive source of energy. The problem with the use of solar energy is not one of insufficient availability, but rather one of finding efficient and economical ways of collecting and storing it so that it will be available when needed. To have energy from the sun available during nights and cloudy days, sufficient solar energy would need to be collected and stored during sunny days. Efficient storage systems are being tested. Sodium thiosulfate pentahydrate is one substance that is being considered as a suitable storage agent. Solar heating can melt it to form a liquid that easily recrystallizes to release energy. Sodium thiosulfate pentahydrate can be easily transported and poured into heat-storage containers, it has a low melting point (48°C) permitting melting at a temperature easily achieved by solar collectors, it is not corrosive, it is inexpensive, and its molar heat of fusion is high enough to collect useful amounts of solar energy per unit of mass.

During this experiment, you will determine the amount of heat energy (enthalpy) that can be stored and subsequently released by the crystallization of sodium thiosulfate pentahydrate. The heat energy (enthalpy) released upon crystallization is referred to as the heat of crystallization or heat of solidification. The heat energy (enthalpy) absorbed during melting is referred to as the heat of fusion or heat of liquefaction. You will study the relationship between heat of fusion and heat of crystallization. You will also determine whether it is easier to directly determine the heat of crystallization or the heat of fusion. How can either the heat of fusion or the heat of crystallization of $Na_2S_2O_3 \cdot 5H_2O$ be determined?

OBJECTIVES

You will heat a small sample of sodium thiosulfate pentahydrate and observe its melting. Based upon your observations, you are to conclude which is spontaneous at room temperature and which is spontaneous at temperatures above the melting point of the material. You will construct a calorimeter using a Styrofoam cup, test its efficiency, and use it to determine the heat of crystallization of sodium thiosulfate pentahydrate. You will need to take many temperature readings, graph your data, extrapolate to determine needed values for ideal (instantaneous) crystallization and heat transfer, and make the necessary calculations in order to express your values in an acceptable form. This experiment provides experience in the determination and use of enthalpy values like those often found in tables.

CONCEPTS

When a solid is slowly heated, it will begin to melt at its melting point (MP). While heat is being supplied to the sample, both solid and liquid will be present and the temperature will not change until all solid is converted to liquid. The heat consumed during the formation of the liquid is known as the heat (enthalpy) of fusion because this process is carried out at constant pressure (barometric pressure) and causes the solid to fuse. When all solid has melted, additional heat will increase the temperature of the now homogeneous melt. When a homogeneous melt at the temperature of its melting point is allowed to lose heat to the surroundings, the melt will begin to solidify. This temperature, known as the freezing point (FP), is equal to the melting point (MP) and will not change as long as liquid and solid coexist. The heat (enthalpy) of crystallization is liberated during the process of the conversion of a melt to a crystalline solid. When all the liquid has solidified, the temperature of the solid will begin to decrease as heat dissipates into the surroundings. To make the heat of fusion and the heat of crystallization independent of the size of the samples and, thus, an intensive property of the substance, the values are typically reported for one mole of the substance (molar heats of fusion or crystallization) or for one gram of a substance (specific heats).

In principle, the specific heat of fusion/crystallization of $Na_2S_2O_3 \cdot 5H_2O$ can be determined if one is able to weigh a sample of the salt, melt it completely at its melting point, let the melt crystallize, and measure the heat given off by the sample during the period between the appearance of the first crystal and the crystallization of the last droplet of liquid. The heat liberated during the crystallization process can be measured by allowing the crystallization process to take place within a calorimeter.

A calorimeter is a well-insulated container that allows a minimum amount of heat energy to escape to the surroundings. In this experiment, a Styrofoam cup will serve as the calorimeter. This type of calorimeter is sometimes referred to as a "coffee cup calorimeter." The calorimeter contains water, the mass of which is exactly known. When the vial with the melt at the freezing point is immersed in the water and crystallization begins, the enthalpy of crystallization is released to the water; the temperature of the water will rise. Under the assumptions of instantaneous crystallization, instantaneous transfer of heat to the water, instantaneous mixing of the water, no heat energy lost to the atmosphere above the calorimeter, and no energy used to warm the calorimeter, the temperature of the water in the calorimeter will rise to a maximum temperature and remain at that temperature. Since the above assumptions cannot be made, it will be necessary to allow the crystallization to conclude, the mixing of the water to be completed, and the calorimeter to be warmed before one obtains cooling data that can be extrapolated back to time of mixing to obtain a temperature value equal to the ideal temperature. You will also need to account for the heat needed to warm the calorimeter and the energy lost to the atmosphere above the calorimeter. For this you need to determine a "heat capacity" for your calorimeter.

To determine the value of the heat capacity (C_{cal}), you place a certain mass of water (m_{rm}) at room temperature (T_{rm}) into the calorimeter to be used. A container with a known mass of water (m_b) is brought to a gentle boil. The boiling water (mass = m_b and temperature = T_b) is poured into the water in the calorimeter. The temperature is measured every minute for a period before and after mixing, and then a plot similar to the one below can be prepared.

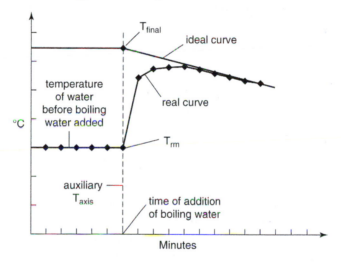

The linear portion of the curve during which cooling is occurring can be extrapolated to the time of mixing and T_{final} read from the new auxiliary axis. The law of conservation of energy (the First Law of Thermodynamics) can be applied to this experiment. From the difference between the temperature of the boiling water (T_b) and the final extrapolated temperature (T_{final}), the mass of the boiling water (m_b), and the specific heat of water (4.18 J g^{-1} degree^{-1}), one can calculate the energy (enthalpy) lost by the boiling water. From the difference between the temperature of the room temperature water (T_{rm}) and the final extrapolated temperature (T_{final}), the mass of the room temperature water (m_{rm}), and the specific heat of water (4.18 J g^{-1} degree^{-1}), one can calculate the energy (enthalpy) gained by the room temperature water. The difference between enthalpy gained by the room temperature water and the enthalpy lost by the boiling water will equal the energy gained by the calorimeter (or the parts of the system other than the water).

To determine the heat of crystallization, a plot similar to the one below can be prepared from experimental data. A vial containing a known mass of molten sodium thiosulfate pentahydrate is placed into a calorimeter, which contains room temperature water (T_{rm}). The sodium thiosulfate pentahydrate crystallizes and the heat released is determined. The temperature is measured every minute before and after the insertion of the vial, and a plot similar to the one below can be prepared.

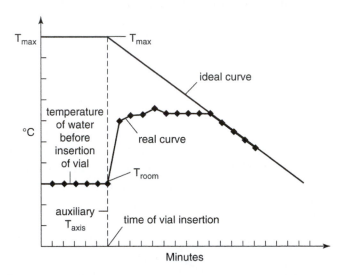

Again, based on the First Law of Thermodynamics: energy cannot be created or destroyed. The energy liberated as heat during the crystallization of the salt must be equal (but of opposite sign) to the energy taken up by the water when its temperature rises to T_{max}. Because the mass of water (m_w) in grams was raised from T_{rm} to T_{max}, an equation that relates the enthalpy gained by the water and the calorimeter to the enthalpy of crystallization can be constructed.

Your calculations will permit a fairly accurate determination of heat of crystallization. A more accurate determination would take into account the heat capacity of the vial, the temperature of the melt at the time of immersion, and a few additional fine points. In this experiment, the temperature of the melt and the vial at the time of immersion are nearly equal to the extrapolated value of T_{max} thus making the amount of heat absorbed by the vial and the salt very small. Thus, we will ignore these factors.

TECHNIQUES

Weighing, heating, and measuring temperature are the techniques used in the laboratory portion of this experiment. A major portion of this experiment deals with the treatment of data. You will need to graph your results and extrapolate the temperature values to obtain an idealized value. You will use your data (some are extrapolated values) to extend our knowledge of a compound that might be considered in constructing solar energy collectors.

ACTIVITIES

You will construct a "coffee-cup calorimeter," determine its efficiency (heat capacity), and use the calorimeter to obtain data needed for the determination of the heat of crystallization of a substance that has potential as an important component in constructing solar energy collectors.

CAUTION

You must wear approved eye protection and exercise proper care in heating and in transferring *hot equipment* and reagents.

PROCEDURES

This experiment requires more than one pair of hands and eyes. Therefore, work with the partner(s) assigned by your instructor. Each of you must keep a complete set of notes, carry out all the calculations, and submit a Report Form.

Observation of Sample

1-1. Obtain a few crystals of sodium thiosulfate pentahydrate in a small test tube. While holding the test tube with a test tube holder, wave it in the flame of a burner. Continue until the salt has melted. Carefully observe the contents of the test tube. Record your observations.

Construction and Use of a Calorimeter

2-1. Obtain an 8-ounce or larger Styrofoam cup. This will serve as your calorimeter. Weigh the cup to the nearest 0.01 g using the top-loader balance.

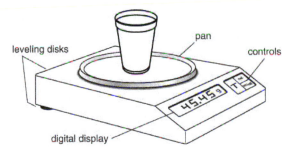

2-2. Using a graduated cylinder, measure 200 mL distilled water into a 250-mL beaker. Set the beaker and water aside for use in a later step (Procedure 3-7).

2-3. Using the graduated cylinder again, measure 100 mL of distilled water into the cup. Weigh the cup and water to the nearest 0.01 g on the top-loader balance.

2-4. Pour 80 mL of distilled water into a 125-mL Erlenmeyer flask. Heat the water to boiling. Adjust the burner to keep the water gently boiling. If the thermometer used in this experiment has limited range and cannot be used to measure the temperature of boiling water, find the boiling point of the water at the barometric pressure in the laboratory from the pressure/boiling point graph on the next page or a table found in the Appendix of your lecture textbook or another reference book.

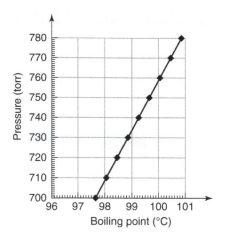

2-5. Prepare a table in your notebook that will allow you to record the temperature readings each minute for 12 or more minutes.

Time (min)	1	2	3	4	5	6	7	8	9	10	11	12
Temp (°C)												

Use a thermometer with marks of 1-degree (or less) to gently stir the water in the calorimeter in a circular motion. Do not let the thermometer drag on the bottom of the cup or rub the sides of the cup. Practice reading the thermometer. If your thermometer has 1-degree markings, you can estimate the reading to tenths of a degree. Stir the water and practice recording the temperature every minute for 4 minutes. Stop stirring to read the thermometer. Read and record all temperatures to the nearest tenth of a degree. ***TIMING IS CRITICAL IN THE NEXT STEPS.*** *Be sure you and your partner have carefully planned your work for efficient operation.*

2-6. Stir the water in the calorimeter. Read and record the temperature of the calorimeter water every minute for 5 minutes. Between the fifth and the sixth minute have your partner remove the flask of boiling water from the ringstand using a towel or tongs to protect fingers.

2-7. Exactly at the start of the sixth minute pour approximately 65 mL of the boiling water into the calorimeter. Continue reading the time and temperature every minute. The temperature will rise quickly to a maximum and then decrease slowly. Keep stirring and recording temperatures every minute until the temperature change per minute is constant for at least 5 minutes. It is very important that you have at least five points in the straight-line segment of the temperature-time graph.

Remove the thermometer from the calorimeter. Let all the water from the stem and bulb of the thermometer drain into the cups. Dry the thermometer with a towel. Determine the mass of the calorimeter and water to 0.01 g on the top-loader balance. Don't empty the calorimeter since it will contain the room-temperature water (or water that is slightly above room temperature) needed in subsequent procedures.

Heat of Crystallization

3-1. Pour 450 mL of distilled water into a 600-mL beaker. Bring the water to a gentle boil.

3-2. Using a top-loader balance, weigh a clean, dry screw-cap vial and the cap to 0.01 g. The screw-cap vial should have a height of 7 cm with a 2.5-cm outside diameter.

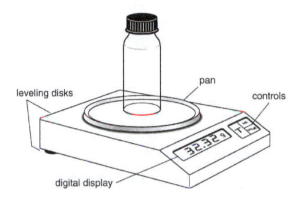

3-3. Fill a 20-cm test tube half full with crystals of $Na_2S_2O_3 \cdot 5H_2O$. Set the test tube containing the sodium thiosulfate into the boiling water in the 600-mL beaker. The crystals will soon begin to melt. You must not allow all the crystals to melt. Time your work to prevent all the crystals from melting and the melt from being heated to higher temperatures.

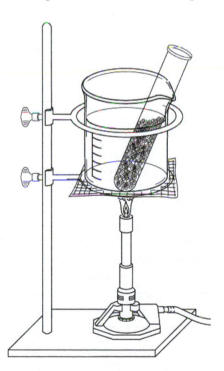

3-4. While the sodium thiosulfate is being heated, carefully push the upper end of the alcohol thermometer (1 degree or smaller divisions) into a split cork stopper until the top of the thermometer has almost reached the top of the stopper. This operation will be easier when the end of the thermometer is moistened with water. Check that it is held tightly in the split and cannot slip out. Attach a clamp to a ringstand and tighten the

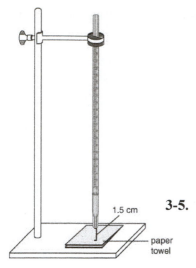

1.5 cm

paper
towel

claws of the clamp sufficiently to prevent the thermometer from slipping. Raise or lower the clamp until the thermometer bulb is suspended approximately 1.5 cm above the base of the ringstand. Place a twice-folded piece of paper towel on the base of the ringstand to serve as insulation.

Procedures 3-6, 3-7, and 3-8 must be completed without any delays between procedures. Plan your work. In your notebook, prepare a time/temperature table (see Procedure 2-5) allowing for entries of up to 15 minutes for the melt data and a second table for up to 40 minutes for the calorimeter data.

3-5. When almost all of the crystals in the test tube have melted, remove the tube from the beaker and pour the melt into the previously weighed vial until the liquid level is just below the neck of the vial. Protect your fingers from the heat.

3-6. Lift the thermometer, slide the vial with the thiosulfate melt into place, and immerse the thermometer bulb into the melt. If the thermometer bulb rests on the bottom of the vial, raise the clamp slightly. Now stir the melt by moving the vial in a circle without lifting it. Be careful to keep the thermometer from hitting the walls or the bottom of the vial. Begin to read and record the temperature of the melt every minute as soon as the temperature has dropped to 48°C.

3-7. Continue recording the temperature every minute. When the temperature reaches 45°C (super-cooled condition), carefully remove the thermometer. Place the thermometer into the 250-mL beaker of room-temperature water until the thermometer registers room temperature, then wipe it dry and begin to stir the calorimeter water with the thermometer. Read and record the temperature of the calorimeter water every minute for 5 minutes on the calorimeter table. This temperature value will be referred to as the temperature of the room temperature water (T_{rm}). Between the additional fifth and sixth minutes of this stirring process have your partner drop one tiny crystal of thiosulfate into the vial and close the vial tightly. At the start of the sixth minute of the stirring process, slide the sealed vial carefully into the calorimeter. The thiosulfate must still be liquid except for the added "seed" crystal. Stir the calorimeter water around the vial as described in Procedure 2-5. Continue to read and record the temperature every minute to the nearest 0.1°C that you can discern. The temperature will slowly rise to a high, stay near the high for a few minutes, and then begin to drop. Continue stirring and recording the temperature. (Be sure you do not unscrew the vial cap during the stirring.)

NOTE: If the salt began to crystallize before the vial was completely immersed in the calorimeter water, you will have to remove the vial, let all water drain back into the calorimeter, tie a string around the vial's neck, and reheat the vial by submerging it in the hot water contained in the 600-mL beaker. Then repeat Procedures 3-6 and 3-7 but this time increase the temperature by 1 or 2 degrees.

3-8. When the cooling rate has remained constant for 7 minutes and the temperature is 2.5 degrees or more below the maximum reading, remove the thermometer and vial, and wipe the vial dry. Carefully unscrew the cap and wipe off any water adhering to the neck of the vial. Reseal the vial and weigh it to the nearest 0.01 g. Record this mass. Reweigh the calorimeter plus water if you suspect it to be different than that obtained at the at the end of Procedure 2-7.

When you have finished, tie a piece of string around the neck of the vial and heat the vial in boiling water until the crystals melt. Pour the molten salt into the specified waste container; then wash and dry the vial. Have your instructor sign your notebook and Report Form.

3-9. Plot your data and determine the temperatures T_{max} and T_{final} by extrapolation. Calculate the heat capacity of the calorimeter and the heat of crystallization of $Na_2S_2O_3 \cdot 5H_2O$. Complete the Report Form.

18   EXPERIMENT 18: HEAT OF CRYSTALLIZATION

Prelab Exercises

1. Give the Problem Statement for this experiment.

2. State in your own words the First Law of Thermodynamics.

3. Define:

heat capacity of the calorimeter

specific heat of crystallization

super-cooled condition

4. What is meant by the phrase "heat gained"?

5. If a thermometer has markings at every degree and the reading is just over 26°C, what temperature should be recorded? Explain your response.

6. Answer any question assigned by your instructor.

Date _____   **Student's Signature** _____

18 EXPERIMENT 18: HEAT OF CRYSTALLIZATION

Report Form

DATA

Barometric pressure _____ torr

Boiling point of water at this pressure (T_b) _____ °C

Observations during the heating and cooling of a sample of sodium thiosulfate pentahydrate:

Mass of empty calorimeter (Procedure 2-1): _____ g

Mass of calorimeter + (~100 mL) cold water (m_{rm}) (Procedure 2-3): _____ g

Temperature of room temperature water (T_{rm}): _____ °C

Mass of calorimeter + room temperature water (m_{rm}) + boiling water (m_b): _____ g

Mass of empty vial and cap (Procedure 3-2): _____ g

Mass of vial and cap and thiosulfate (m_{salt}) (Procedure 3-8): _____ g

Date _____ **Instructor's Signature** _____

ANALYSIS

Hot and cold water experiment

Mass of cold water (m_{rm}) (Procedures 2-1 and 2-2): _____ g

Mass of boiling water (m_b) (Procedures 2-3 and 2-7): _____ g

Maximum temperature for mixed boiling and room temperature water
(T_{final}, extrapolated from attached graph of data
obtained during Procedure 2-7): _____ °C

Specific heat of water (C_w) (See Concepts section): _____ J g^{-1} degree^{-1}

An equation that relates energy gained by room temperature water to m_{rm}, C_w, T_{final}, and T_{rm} is:

_____

The numerical value for the energy gained by the room temperature water calculated using the above

equation: _____

An equation that relates the energy lost by the boiling water to m_b, C_w, T_{final}, and T_b is:

_____

The numerical value for the energy lost by the boiling water calculated using the above

equation: _____

An equation that relates energy gained by the calorimeter to the energy gained by room temperature water and the energy lost by boiling water is:

_____

The numerical value for the energy gained by the calorimeter using the above equation:

_____

If the above value for energy gained by the calorimeter is defined as heat capacity calculated for your calorimeter (C_{cal}) times change in temperature of the calorimeter, what is the calculated value for your

calorimeter: _____ J/°C?

Cold water and melted sodium thiosulfate experiment

Temperature of room temperature water (Procedure 3-7): T_{rm} _____ °C

Mass of room temperature water (m_{rm}) + boiling water (m_b) (Procedures 2-3 and 2-7):

m_w _____ g

Temperature of water after crystallization of thiosulfate
(T_{max}, extrapolated from data obtained during Procedure 3-9) _____ °C

Energy gained by water: (equation form) _____

(numerical value) _____ J

Using the heat capacity determined above for the calorimeter, the energy gained by calorimeter is:

(equation form) _____

(numerical value) _____ J

Heat released by crystallization of sample is equal to (but opposite in sign) the sum of the two values

above: _____ J

Heat released by crystallization of one gram of sample:

(specific heat of crystallization) _____ J/g

Heat released by crystallization of one mole of sample:

(molar heat of crystallization) _____ kJ/mol

POSTLAB QUESTIONS

1. Compare the specific heat of crystallization of sodium thiosulfate pentahydrate with the corresponding data for water (-334 J/g) and sodium chloride (-519 J/g). Is sodium thiosulfate a better material to collect solar energy per gram of material than water, than sodium chloride? Why?

2. Explain why the specific heat of crystallization is often the value of choice when comparing materials for the design of a solar energy collection system.

3. Besides having a low melting point and a high specific heat of crystallization, what other properties should be considered when researching appropriate compounds for use in a solar collector?

4. Is crystallization a reversible process? If so, what is the reverse process called and how does its enthalpy relate to that for crystallization?

5. Explain why it is necessary to write a negative sign in an expression that equates the energy lost by the more energetic body in a calorimeter experiment as being equal to the energy gained by the less energetic bodies.

6. In this experiment, why was it important that the seed crystal be very small?

7. What properties are desirable when selecting a calorimeter? Could a thermos bottle or cooler be used as a calorimeter? Explain your reasoning.

8. Give an example that relates the First Law of Thermodynamics to an observation or event that you have observed outside of class.

Date _____ **Student's Signature** _____

Enthalpy of Reactions

A Guided Inquiry Experiment

INTRODUCTION

In the ideal world, one could design an ideal calorimeter that would completely insulate the reacting solutions from the surroundings and, therefore, allow all the heat liberated by the reaction to be used to raise the temperature of the aqueous solution in the calorimeter. From the temperature of the solution before the reaction began and the temperature of the solution after the reaction was complete, the temperature change could be measured. Calculations based on the First Law of Thermodynamics (the heat generated by the reaction equals the heat gained by the solution) could be performed in that ideal world without consideration of energy used to warm the calorimeter and its surroundings. For example, one could make the following calculations:

$$\Delta H_{rxn} = - \Delta H_{soln}$$

The ideal calorimeter does not exist. The calorimeter used in this experiment does not completely prevent the loss of heat to the surroundings. These heat losses can be taken into account through a heat capacity value for each specific calorimeter (C_{cal}). The heat capacity of the calorimeter is the heat taken up by the calorimeter and/or lost to the surroundings per degree temperature rise. The value of the heat capacity of the calorimeter can be determined experimentally, as was done in Experiment 17, by mixing in a calorimeter a known mass of water (m_{rm}) at room temperature (T_{rm}) with a known mass (m_b) and temperature (T_b) of boiling water and determining the temperature of the resulting mixture (T_{final}). Fortunately in this experiment, you will be investigating the heats of a number of reactions using the same calorimeter that is very efficient. You will use similar quantities of solution in each determination, and you will be primarily interested in differences in heats of the reactions. Therefore, the heat absorbed by the calorimeter will be small and nearly equal in each determination. During this experiment, the heat taken up by the calorimeter and the heat lost to the surroundings will be ignored.

For each reaction in this experiment, all the quantities on the right-hand side of this equation should be known at the end of the experiment and the enthalpy of the reaction can be calculated as:

$$\Delta H_{rxn} = -[m_s C_s (T_{arxn} - T_{brxn})]$$

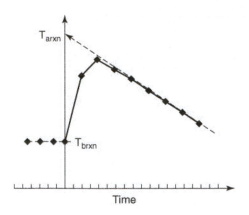

ΔH_{rxn}: enthalpy of the reaction performed with known quantities of reagents

m_s: mass of solution in the calorimeter (density of solution is 1.02 g mL^{-1})

C_s: specific heat of the solution (3.97 J g^{-1} degree^{-1} for the solutions that will be used in this experiment)

T_{brxn}: temperature of solution before reaction (extrapolated to time of mixing). This is often referred to as the initial temperature.

T_{arxn}: temperature of solution after instantaneous reaction (extrapolated to time of mixing). This is also referred to as the final or maximum temperature.

OBJECTIVES

In this experiment, you will work with a coffee cup calorimeter. You will use the calorimeter to determine heats of specific reactions. The relationships between the heats of these reactions will be further investigated.

CONCEPTS

This experiment uses the concepts of stoichiometry, sequential reactions, and energy transfer. You will demonstrate the usefulness of the simple calorimeter.

TECHNIQUES

Organization of the collecting and recording of data is most important during this experiment. You will need to record numerous time and temperature readings.

ACTIVITIES

In this experiment, you will obtain the needed equipment, run the needed experiments, plot the data obtained, determine the energy transferred, write thermochemical equations, and use your results to investigate the relationships between various reactions and the heats of those reactions.

| CAUTION | |

Use approved eye protection. Be aware of chemicals and hot equipment. Follow the details provided by your instructor.

PROCEDURES

Equipment Construction

1-1. For a calorimeter to be used throughout this experiment obtain an 8-oz or larger Styrofoam cup. Weigh the cup to the nearest 0.01 g on the top-loader balance. Record the mass.

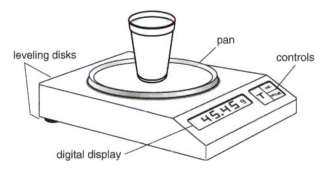

Enthalpy of NaOH(s) →
NaOH(aq)

2-1. With a graduated cylinder measure 100 mL of distilled water. Pour the 100 mL of distilled water into the calorimeter. Take the calorimeter with the water to the top-loader balance and weigh it to the nearest 0.01 g. Record the mass.

2-2. Using the top-loader balance, weigh out 1.5 to 2.0 g of NaOH. Record the exact mass to the nearest 0.01 g.

2-3. Suspend a thermometer in the water in the calorimeter; take five temperature readings at 20-second intervals; and at the sixth reading, add the solid NaOH. Swirl the solution gently and take at least 14 more temperature readings at 20-second intervals, making a note when all of the NaOH dissolves. If 14 additional readings is not 10 readings past the point where all the NaOH has dissolved, continue to collect data until you have at least 10 readings past when the last solid NaOH dissolved.

2-4. Empty the solution from the calorimeter into the sink. Rinse the calorimeter with distilled water. Dry the calorimeter inside and outside.

Enthalpy of HCl(aq) +
NaOH(s)

3-1. Take a clean and dry graduated cylinder (50 mL or 100 mL) to the 1._ _ *M* hydrochloric acid storage bottle. (You may need to prepare this by dilution of a slightly more concentrated solution.) Cautiously pour the acid into the graduated cylinder until the liquid level is just above the 50-mL mark. Take the graduated cylinder with the dilute acid in it to your desk. With a clean dropper, remove some of the acid until the meniscus touches the 50-mL mark. Rinse the dropper and pour excess acid down the sink with large amounts of water. Pour the 50.0 mL of acid into the calorimeter assembly. Make sure that all 50.0 mL of the dilute acid is transferred. Rinse the graduated cylinder and invert it to dry.

3-2. Use the top-loader balance and weigh out 1.50 to 2.00 g of NaOH. Record the exact mass to the nearest 0.01 g.

3-3. Suspend a thermometer in the acid in the calorimeter; take five temperature readings at 20-second intervals; and at the sixth reading, add the solid NaOH. Swirl the solution gently and take at least 14 more temperature readings at 20-second intervals. If that number of readings is not 10 past the point where all the NaOH has dissolved, continue to collect data until you have at least 10 readings past that point.

3-4. Empty the solution from the calorimeter into the sink. Rinse the calorimeter with distilled water. Dry the calorimeter inside and outside.

Enthalpy of HCl(aq) + NaOH(aq)

4-1. Repeat Procedure 3-1 to obtain 50 mL of 1.__ _ M HCl.

4-2. Into another clean and dry graduated cylinder, pour 50 mL of 1.__ _ M sodium hydroxide solution as described above. (You may need to prepare this by dilution of a slightly more concentrated solution.) Cautiously pour the base into the graduated cylinder until the liquid level is just above the 50-mL mark. Take the graduated cylinder with the dilute base in it to your desk. Use a clean dropper to remove some of the base until the meniscus touches the 50-mL mark. Rinse the dropper and wash the excess base down the sink with large amounts of water. The hydrochloric acid and the sodium hydroxide solution must have the same temperature. This condition is fulfilled when the storage bottles have been in the laboratory for several hours.

4-3. Suspend a thermometer in the acid in the calorimeter; take five temperature readings at 20-second intervals; and at the sixth reading, add the base. Swirl the solution gently and take 14 more temperature readings at 20-second intervals.

4-4. Empty the solution from the calorimeter into the sink. Rinse the calorimeter with distilled water. Dry the calorimeter inside and outside.

Enthalpy of Acetic Acid(aq) + NaOH(aq

5-1. Repeat Procedures 4-1 through 4-4 using 50 mL of 1.__ _ M acetic acid in the place of the HCl solution. Remember that acetic acid can be written as CH_3COOH or $HC_2H_3O_2$.

5-2. Have your instructor sign your notebook and Report Form.

19   EXPERIMENT 19: ENTHALPY OF REACTIONS

Prelab Exercises

1. Write an example of formula unit equations and the net ionic equations for the neutralization of a strong, binary acid with a strong, dibasic base; a weak acid with a strong base; and a weak base with a strong, diprotic acid. (Use examples that include only soluble acids, bases, and salts.)

2. Write the net ionic equation for the neutralization of aqueous hydrochloric acid with an aqueous solution of sodium hydroxide.

3. Define the First Law of Thermodynamics and describe how it applies to this experiment.

4. For an exothermic reaction carried out in a calorimetric experiment, the following temperature data were collected at 1-minute intervals with the exothermic reaction beginning at the fourth temperature reading: 22.08, 22.18, 21.98, 22.08, 26.78, 29.18, 30.58, 30.48, 30.28, 30.08, 29.88, 29.68, 29.48, 29.38. Illustrate how the temperature expected for instantaneous reaction can be determined by extrapolation. Attach a sheet of graph paper.

5. How many moles of hydronium ions were converted to water when 80.0 mL of 1.1 *M* hydrochloric acid were neutralized by sodium hydroxide? (Note: The terms hydronium ions and aqueous hydrogen ions are often used interchangeably.)

Date _____ **Student's Signature** _____

19 EXPERIMENT 19: ENTHALPY OF REACTIONS

Report Form

DATA

Data for the determination of the enthalpy of reaction for NaOH(s) → NaOH(aq)

Mass of empty calorimeter: _____ g

Mass of calorimeter and room temp. water: _____ g

Mass of room temp. water: _____ g

Mass of NaOH: _____ g

Moles of NaOH: _____ mol

Total mass of solution (m_s): _____ g [$m_{water} + m_{NaOH}$]

Time/Temperature Data:

0 s	°C	80 s	°C	160 s	°C	240 s	°C	320 s	°C
20 s	°C	100 s	°C	180 s	°C	260 s	°C	340 s	°C
40 s	°C	120 s	°C	200 s	°C	280 s	°C	360 s	°C
60 s	°C	140 s	°C	220 s	°C	300 s	°C	380 s	°C

Determination of the enthalpy of reaction for NaOH(s) → NaOH(aq)

Attach a plot of the time/temperature data recorded above and attach all calculations. (See details for graphing data in the Common Procedures and Concepts Section at the end of this manual.)

Temperature of room temperature water at time of mixing (T_{brxn}): _____ °C
(Determined by extrapolation and indicated on the attached plot.)

Extrapolated maximum temperature of mixture (T_{arxn}): _____ °C

$\Delta T_{reaction}$: _____ °C ($T_{arxn} - T_{brxn}$)

Heat energy gained by the solution: _____ J or _____ kJ
[$m_s C_s(T_{arxn} - T_{brxn})$] and ($C_s = 3.97$ J g^{-1} degree^{-1})

What is the value you obtained for ΔH_{rxn} for the moles of NaOH used?

_____ J or _____ kJ

ΔH formation of one mole of NaOH: _____ kJ mol^{-1}
(ΔH_{rxn} divided by number of moles of rxn.)

Data for the determination of the enthalpy of reaction for HCl(aq) + NaOH(s)

Volume of 1._ _ M HCl: _____ mL (3 significant figures)

Mass of solution (mHCl): _____ g (vol. in mL $\times$ 1.02 g mL^{-1})

Moles of HCl: _____ mol (1._ _ $M \times V_{HCl}/1000$) (3 significant figures)

Mass of NaOH: _____ g

Moles of NaOH: _____ mol

Total mass of solution (m_s): _____ g [$m_{HCl} + m_{NaOH}$]

Time/Temperature Data:

0 s	°C	80 s	°C	160 s	°C	240 s	°C	320 s	°C
20 s	°C	100 s	°C	180 s	°C	260 s	°C	340 s	°C
40 s	°C	120 s	°C	200 s	°C	280 s	°C	360 s	°C
60 s	°C	140 s	°C	220 s	°C	300 s	°C	380 s	°C

Determination of the enthalpy of reaction for HCl(aq) + NaOH(s)

Attach a plot of the time/temperature data recorded above and attach all calculations.

Temperature of HCl solution at time of mixing (T_{brxn}): _____ °C
(Determined by extrapolation and indicated on the attached plot.)

Extrapolated Temperature (max) of solution(s) at time of mixing for instantaneous neutralization

(T_{arxn}): _____ °C

Heat energy gained by the solution: _____ J or _____ kJ
[$m_s C_s(T_{arxn} - T_{brxn})$] and ($C_s = 3.97$ J g^{-1} degree^{-1})

Heat energy of reaction (ΔH_{rxn}): _____ J or _____ kJ
(Remember that this is the heat for the number of moles of water formed in the reaction, which is equal to the smaller of the number of moles of HCl or moles of NaOH available to react.)

ΔH for one mole of reaction: _____ kJ mol^{-1}
(ΔH_{rxn} divided by number of moles of rxn)

Data for the determination of the enthalpy of reaction for HCl(aq) + NaOH(aq)

If the following data is not your own, indicate which group provided it. (Initials are sufficient.)

Volume of 1._ _ *M* HCl: _____ mL (3 significant figures)

Mass of solution (m_{HCl}): _____ g (vol. in mL × 1.02 g mL^{-1})

Moles of HCl: _____ mol (1._ _ *M* × V_{HCl}/1000) (3 significant figures)

Volume of 1._ _ *M* NaOH: _____ mL (3 significant figures)

Mass of solution (m_{NaOH}): _____ g (vol. in mL × 1.02 g mL^{-1})

Moles of NaOH: _____ mol (1._ _ *M* × V_{NaOH}/1000) (3 significant figures)

Total volume of solution: _____ mL

Total mass of solution (ms): _____ g [m_{HCl} + m_{NaOH}]

Time/Temperature Data:

0 s	°C	80 s	°C	160 s	°C	240 s	°C	320 s	°C
20 s	°C	100 s	°C	180 s	°C	260 s	°C	340 s	°C
40 s	°C	120 s	°C	200 s	°C	280 s	°C	360 s	°C
60 s	°C	140 s	°C	220 s	°C	300 s	°C	380 s	°C

Determination of the enthalpy of reaction for HCl(aq) + NaOH(aq)

Attach a plot of the time/temperature data recorded above and attach all calculations.

Temperature of HCl solution at time of mixing (T_{brxn}): _____ °C
(Determined by extrapolation and indicated on the attached plot.)

Extrapolated Temperature (max) of solution(s) at time of mixing for instantaneous neutralization

(T_{arxn}): _____ °C

Heat energy gained by the solution: _____ J or _____ kJ
[$m_s C_s (T_{arxn} - T_{brxn})$] and ($C_s$ = 3.97 J g^{-1} degree^{-1})

Heat energy liberated by the reaction (ΔH_{rxn}): _____ J or _____ kJ
(Remember that this is the heat for the number of moles of water formed in the reaction, which is equal to the smaller of the number of moles of HCl or moles of NaOH available to react.)

ΔH for one mole of reaction: _____ kJ mol^{-1}
(ΔH_{rxn} divided by number of moles of rxn.)

Data for the determination of the enthalpy of reaction for CH$_3$COOH(aq) + NaOH(aq)

Volume of 1._ _ *M* CH$_3$COOH: _____ mL (3 significant figures)

Mass of solution (m_{acetic}): _____ g (vol. in mL $\times$ 1.02 g mL^{-1})

Moles of CH_3COOH: _____ mol (1._ _ $M \times V_{acetic}$/1000) (3 significant figures)

Volume of 1._ _ M NaOH: _____ mL (3 significant figures)

Mass of solution (mNaOH): _____ g (vol. in mL 1.02 g mL^{-1})

Time/Temperature Data:

0 s	°C	80 s	°C	160 s	°C	240 s	°C	320 s	°C
20 s	°C	100 s	°C	180 s	°C	260 s	°C	340 s	°C
40 s	°C	120 s	°C	200 s	°C	280 s	°C	360 s	°C
60 s	°C	140 s	°C	220 s	°C	300 s	°C	380 s	°C

Determination of the enthalpy of reaction for $CH_3COOH(aq)$ + $NaOH(aq)$

Attach a plot of the time/temperature data recorded above and attach all calculations.

Temperature of HCl solution at time of mixing (T_{brxn}): _____ °C

Extrapolated Temperature (max) of solution(s) at time of mixing for instantaneous neutralization

(T_{arxn}):_____ °C

Heat energy gained by the solution: _____ J or _____ kJ
[$m_sC_s(T_{arxn} - T_{brxn})$] ($C_s$ = 3.97 J g^{-1} degree^{-1})

Heat energy liberated by the reaction (ΔH_{rxn}): _____ J or _____ kJ
(Remember that this is the heat for the number of moles of water formed in the reaction, which is equal to the smaller of the number of moles of acetic acid or moles of NaOH available to react.)

ΔH for one mole of reaction: _____ kJ mol^{-1}
(ΔH_{rxn} divided by number of moles of rxn.)

CLASS DATA

(A) heat per mole rxn for **NaOH(s) $\rightarrow$ NaOH(aq).**

_____ _____ _____ _____ _____

_____ _____ _____ _____ _____

(B) heat per mole rxn for **HCl(aq) + NaOH(s).**

_____ _____ _____ _____ _____

_____ _____ _____ _____ _____

(C) heat per mole rxn for **HCl(aq) + NaOH(aq).**

_____ _____ _____ _____ _____ _____

_____ _____ _____ _____ _____ _____

(D) heat per mole rxn for **CH₃COOH(aq) + NaOH(aq).**

_____ _____ _____ _____ _____ _____

_____ _____ _____ _____ _____ _____

Date _____ **Instructor's Signature** _____

ANALYSIS

1. Is the heat per mole rxn for each reaction above similar? How does your heat per mole rxn compare to those from your class?

2. What is the average heat per mole rxn for each of the four reactions above? If you omit any data, explain why you omitted a particular data point.

    (A) _____ (kJ/mol rxn) for **NaOH(s) → NaOH(aq).**

    (B) _____ (kJ/mol rxn) for **HCl(aq) + NaOH(s).**

    (C) _____ kJ/mol rxn) for **HCl(aq) + NaOH(aq).**

    (D) _____ (kJ/mol rxn) for **CH₃COOH(aq) + NaOH(aq).**

3. Write out the net ionic equations for the reactions of **NaOH(s) → NaOH(aq), HCl(aq) + NaOH(s),** and **HCl(aq) + NaOH(aq).**

a. What is the relationship among the chemical equations? Can any two equations be combined to form the third?

b. Are these reactions exothermic or endothermic? If an exothermic reaction is reversed, will it be exothermic or endothermic? Explain.

c. Which reaction had the highest average heat per mole rxn? Which had the smallest? Which was the middle?

d. How do the relationships among the chemical equations in (a) compare with the relationships among the heats of reactions?

e. Draw a particle view of these three reactions showing the relationship between their equations and heats of reaction in kJ/mol.

4. Write out the net ionic equation for the reaction of $CH_3COOH(aq) + NaOH(aq)$.

5. Explain how you could find the heat of reaction in kJ/mol rxn for the ionization of acetic acid using the values you have found in question 2.

$$CH_3COOH(aq) \rightarrow H^+(aq) + CH_3COO^-(aq)$$

6. Find the numeric value for the heat of reaction in kJ/mol rxn for the ionization of acetic acid using the values in question 2.

7. Why was a calorimeter constant not calculated?

POSTLAB QUESTIONS

1. Do the signs of the values in question 2 of the Analysis section support what you should also conclude about spontaneity if the entropy values times temperature are small compared to these values?

2. Is the heat/mol rxn value for the reaction of $H^+ + OH^-$ any different than the value for $HCl(aq) + NaOH(aq)$? Explain.

3. Is the heat/mol rxn value for the reaction of H^+ + OH^- any different than the value for **HCl(aq)** + **NaOH(s)**? Explain.

4. Is the heat/mol rxn value for the reaction of CH_3COOH + OH^- any different than the value **CH₃COOH(aq)** + **NaOH(aq)**? Explain.

5. Define the terms exothermic and endothermic from the viewpoint of the reactants.

6. Define Hess' Law.

7. In the following cycle, the enthalpies can be measured for all reactions (ΔH_{AB}, ΔH_{BC}, ΔH_{AE}, ΔH_{ED}) except the one shown as a dashed arrow.

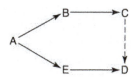

   Show that the molar enthalpy of reaction ΔH_{CD}, step C → D, can be expressed in terms of the other steps.

Date _____ Student's Signature _____

The Kinetics of the Decomposition of Hydrogen Peroxide

A Guided Inquiry Experiment

INTRODUCTION

Rates of reactions (kinetics) are important for many reasons. "How fast will a food item spoil and how can we change that rate?" In this experiment, we study how kinetics might be investigated and what we can conclude from the study of kinetics.

How is the rate of the decomposition measured? In the case of the decomposition of hydrogen peroxide, the moles of dry gas produced can be measured at various times after the reaction has been initiated. Experimentally, it is easier to measure the volume of gas evolved as a function of time. The volumes and moles are related to each other via the Ideal Gas Equation. As the reaction proceeds, the hydrogen peroxide concentration in the solution will decrease and the volume of gaseous product formed per unit time (the rate) will decrease. When all the hydrogen peroxide is decomposed, its concentration is zero, and no more product is formed.

The rate of a reaction may be expressed using a rate-law expression. A rate-law expression contains a lot more information about the reaction than just the rate. The generalized rate-law expression for the iodide catalyzed decomposition of hydrogen peroxide is:

$$R = \text{rate} = k[H_2O_2]^n [I^-]^m$$

The values for n and m must be determined experimentally. n and m are called the "orders" of the reaction with respect to hydrogen peroxide and iodide. The sum of the exponents is the overall order of the reaction. The rate of the reaction is determined by the slowest step in a multistep sequence of reactions (mechanism). k is the rate constant.

Because of experimental errors, the exponents experimentally determined for a rate-law expression are rarely integers. Although fractional orders of reactions are known, in this experiment the exponents are to be rounded to zero or the nearest whole number. With the order of the reaction with respect to H_2O_2 and I^- determined, chemically-based intuition will be used to propose a "mechanism" for the decomposition of hydrogen peroxide that agrees with your rate-law expression.

OBJECTIVES

During this experiment, you will determine the products of the reaction that takes place during the decomposition of hydrogen peroxide as well as investigate the effect of changing concentrations and catalysts on the rate of reaction. You will write a rate-law expression for the iodide and hydrogen peroxide reaction and will determine the values of the exponents and the rate constant in the rate-law expression. Based upon the values that you obtain for the rate-law expression, you are to propose a mechanism.

CONCEPTS

This experiment uses stoichiometric concepts, gas collection equipment, and extensive data handling. The two reactions that you will study are both the decomposition of hydrogen peroxide. The first involves the catalyst iodide ion and the second the enzyme catalase obtained from raw vegetable extract. You will utilize your knowledge of reaction rates and rate-law expressions.

TECHNIQUES

The apparatus and many of the techniques used are also found in Experiment 14. Preparation of solutions, assembly of equipment, and determination of mass, temperature, and pressure are encountered in both experiments.

ACTIVITIES

You will analyze the gas produced during the decomposition of a hydrogen peroxide solution, measure the volume of gas produced as a function of time, run experiments with varying concentrations of reactants and catalysts, and discuss reaction kinetics.

SAFETY

Wear eye protection at all times. Be very careful when working with glassware, particularly when tightening stoppers and slipping rubber tubing on glass tubes. Excessive force in these operations might lead to breakage of the glass and painful cuts.

PROCEDURES

Preparing Solutions

1-1. Prepare 200 mL of a 0.15 *M* solution of KI.

The Apparatus

2-1. Position a buret clamp about 30 cm above the base of a ringstand so that the arm of the clamp extends over the base. Near the top of the ringstand attach a small split-ring clamp at approximately a 45-degree angle to the first clamp. Also near the top of the ringstand, position a clamp that is suitable for holding a 125-mL flask.

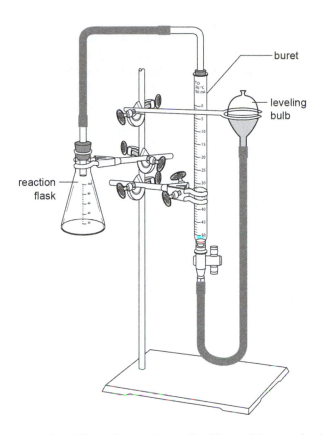

2-2. Attach a 60-cm long piece of rubber tubing to the lower end of a 50-mL buret. Clamp the buret securely. Obtain a No. 00 one-hole rubber stopper that holds a short piece of glass tube with a 90 degrees or smaller bend and a one-hole stopper that will fit (stopper) a 125-mL Erlenmeyer flask. There must be a short piece of straight glass tube in the stopper. Fasten the 125-mL flask securely to the ringstand. Connect the flask and the top of the buret with the 60-cm long rubber tubing.

CAUTION

Do not exert too much pressure on the stopper. You might break the neck of the flask and cut yourself.

2-3. Position a leveling bulb in the split ring on the ringstand. Connect the rubber tubing attached to the bottom of the buret to a leveling bulb.

Studying the Reaction

3-1. You will investigate the reaction(s) that occur when hydrogen peroxide solution reacts with potassium iodide.

Place about 30 drops of 3% H_2O_2 solution in each of two 18 × 150-mm test tubes. Stand the test tubes in a test tube rack or beaker. Into one of the tubes, add an amount of solid KI solution that is sufficient to cover the tip of a spatula. Observe the reaction. Determine the identity of the gas, if any, produced. If a gas is produced, it will be water vapors, hydrogen, or oxygen. Consult the Common Procedures and Concepts Section at the back of this manual for possible tests for these gases.

Use a test tube holder to hold the test tube if you need to heat the contents of the test tube. Stoppers, delivery tubes, etc., will be available if you wish to test for hydrogen. Other materials will be available to test for water vapor and oxygen. REMEMBER: Hydrogen can produce a small explosion capable of launching stoppers, etc.

3-2. Record your results and observations, including those of the undisturbed test tube.

Running of Experiment

4-1. Obtain and record the barometric pressure and temperature in the laboratory. *(You will need this information to determine from the table below the vapor pressure of water used during this experiment.)* Remove the stopper from the Erlenmeyer flask. Pour 125 mL of distilled water at room temperature into the leveling bulb. Lower and raise the bulb several times to expel all air from the tubing. When you can no longer observe bubbles rising to the surface, restore the bulb in its original position.

Vapor Pressure of Water

°C	16	17	18	19	20	21	22	23	24	25	26	27	28	29	30	31
Torr	13.6	14.5	15.5	16.5	17.5	18.7	19.8	21.1	22.4	23.8	25.2	26.7	28.3	30.0	31.8	33.7

4-2. Fill a clean and dry 25-mL graduated cylinder with the 0.15 M KI solution to the 20-mL mark. Pour the 20 mL of KI solution into the 125-mL Erlenmeyer flask. Fill the same graduated cylinder with distilled water at room temperature to the 20-mL mark. Add this water to the flask.

4-3. Tightly stopper the 125-mL Erlenmeyer flask. Check the stopper in the buret for a tight fit. Lower the leveling bulb to a position approximately 10 cm below its original position. Observe the water level in the buret. The water level in the buret will drop and should stop at a level considerably above the level in the bulb. If the water level in the buret keeps dropping slowly, the system has a leak. Check the stoppers and the hose connections. If a joint leaks, try to make it airtight by applying distilled water to the joint rather than using excessive force.

Plan the next step carefully. Ask your partner or neighbor at the laboratory bench to assist you during the measurement of the evolved oxygen.

4-4. Unstopper the 125-mL Erlenmeyer flask; quickly add the 10 mL 3% (0.90 M) H_2O_2 solution at once and immediately stopper the flask tightly. This moment is the beginning of the experiment. Record the time of this moment as read by your assistant. Remove the Erlenmeyer flask quickly from its clamp and swirl the solution in the flask gently. **Constant swirling is required for good results.** Record the volume reading in the buret (with the water level in the leveling bulb exactly at the same height as in the buret) every 20 seconds as announced by your assistant. Continue swirling the flask and recording volumes until approximately 20 mL of gas have been released.

4-5. Place the leveling bulb in its clamp. Cautiously open the stopper in the Erlenmeyer flask. Remove the flask from the clamp, dispose of reagents as directed by your instructor, and thoroughly rinse the flask with distilled water. Dry the inside of the flask with a towel.

4-6. Repeat Procedures 4-2 through 4-5 three more times with the following solutions in the Erlenmeyer flask:

- 20 mL 0.15 M KI, 10 mL H_2O, 20 mL 3% H_2O_2
- 40 mL 0.15 M KI, no water, 10 mL 3% H_2O_2
- 5 mL vegetable extract, 25 mL H_2O, 20 mL 3% H_2O_2

Record the volume every 5 seconds for the vegetable extract.

4-7. Dispose of reagents as directed by your instructor. Leave the equipment clean and ready for the next student to use. Have your instructor sign your notebook and Report Form.

| 20 | **EXPERIMENT 20: THE KINETICS OF THE DECOMPOSITION OF HYDROGEN PEROXIDE** |

Prelab Exercises

1. Using logical compounds, write an equation for the decomposition of hydrogen peroxide to produce hydrogen if such an equation is possible.

2. Using logical compounds, write an equation for the decomposition of hydrogen peroxide to produce oxygen if such an equation is possible.

3. Calculate the mass of KI needed to prepare 200 mL of a 0.15 M solution of KI. Record this question and answer in your lab notebook.

4. A solution is prepared by mixing 10 mL 0.15 M KI, 25 mL of distilled water, and 30 mL 0.90 M H_2O_2. Calculate the concentrations of each of these substances in the final solution before any reaction takes place.

5. During the decomposition of hydrogen peroxide, is it possible to determine the moles of dry gas being obtained using the following equation: $(P_{total} - P_{H_2O})V_{total} = n_{new\ gas} RT$? Explain.

6. The following data were collected in an experiment in which hydrogen peroxide was decomposed to water and unknown gas. The data are given as **mol unknown gas evolved/time** in seconds elapsed since the beginning of the experiment:

    0.0000625/15   0.000125/30   0.000188/40   0.000250/65   0.000313/90

                        0.000375/110   0.000438/130   0.000500/150   0.000563/170

    Plot these data, calculate the slope of the best straight line through zero (you may want to use the least squares program in a calculator), and give the numerical value and the units for the rate of formation of unknown gas.

Date _____   **Student's Signature** _____

20 EXPERIMENT 20: THE KINETICS OF THE DECOMPOSITION OF HYDROGEN PEROXIDE

Report Form

DATA

Observations of hydrogen peroxide test tube with addition of KI:

Results for test of gas produced:

Observations of undisturbed test tube:

Barometric pressure: _____

Laboratory temperature: _____

Partial pressure of water: _____

Mass of KI: _____

Total volume of solution: _____

Kinetics Data: Run #1 (20 mL KI, 20 mL H_2O, 10 mL H_2O_2)

Time	Volume	Time	Volume	Time	Volume

Kinetics Data: Run #2 (20 mL KI, 10 mL H_2O, 20 mL H_2O_2)

Time	Volume	Time	Volume	Time	Volume

Kinetics Data: Run #3 (40 mL KI, 0 mL H_2O, 10 mL H_2O_2)

Time	Volume	Time	Volume	Time	Volume

Kinetics Data: Run #4 (5 mL extract, 25 mL H_2O, 20 mL H_2O_2)

Time	Volume	Time	Volume	Time	Volume

Barometric pressure: _____   Laboratory temperature: _____

Partial pressure of water: _____

Date _____   **Instructor's Signature** _____

ANALYSIS

Proposed reaction:

Moles of gas:

[Use the equation $(P_{total} - P_{H_2O})(V_{total} - V_{init.}) = n_{new\ gas} RT$ to determine the moles of new gas present at each of the times during your kinetics determinations.]

Kinetics Data: Run #1 (20 mL KI, 20 mL H_2O, 10 mL H_2O_2)

Time	Volume	Time	Volume	Time	Volume

Kinetics Data: Run #2 (20 mL KI, 10 mL H_2O, 20 mL H_2O_2)

Time	Volume	Time	Volume	Time	Volume

Kinetics Data: Run #3 (40 ml KI, 0 ml H_2O, 10 ml H_2O_2)

Time	Volume	Time	Volume	Time	Volume

Kinetics Data: Run #4 (5 mL extract, 25 mL H_2O, 20 mL H_2O_2)

Time	Volume	Time	Volume	Time	Volume

Attach a plot of each of the sets of kinetics data listed above.

Draw the best straight line for only the first several data points. Find the initial rate (slope of the straight line at t = zero) for each reaction in terms of moles of gas produced per unit time.

The initial rate for each reaction (slope of each plot) is:

Run #1 _____ (mol/min)   Run #2 _____ (mol/min)

Run #3 _____ (mol/min)   Run #4 _____ (mol/min)

Initial Concentrations:

Run #1 (20 mL KI, 20 mL H_2O, 10 mL H_2O_2) = _____ M KI, _____ M H_2O_2
                                 (3% H_2O = 0.9 mol/L)

Run #2 (20 mL KI, 10 mL H_2O, 20 mL H_2O_2) = _____ M KI, _____ M H_2O_2

Run #3 (40 mL KI, 0 mL H_2O, 10 mL H_2O_2) = _____ M KI, _____ M H_2O_2

Run #4 (5 mL extract, 25 mL H_2O, 20 mL H_2O_2) = _____ M extract, _____ M H_2O_2
                                 (undiluted extract = 0.01 mol/L)

Comparisons:

What relationship do you see between the rates and concentrations of Run #1 and Run #2? How did the concentration of hydrogen peroxide affect the rate?

What relationship do you see between the rates and concentrations of Run #1 and Run #3? How did the concentration of KI affect the rate?

Describe any generalizations concerning the rate and concentrations of KI and hydrogen peroxide.

What relationship do you see between the rates and concentrations of Run #2 and Run #4?

Rate Law:

The exponents in the rate-law expression can be determined by writing the expression with the known data for rate and concentrations from two runs. That will give two equations in which k, m, and n are unknown.

$$R = k[H_2O_2]^n[I^-]^m \quad \text{and} \quad R' = k[H_2O_2']^n[I^{-\prime}]^m$$

If only one of the concentrations varies and if k is a constant, which it will be if the temperature does not change between runs, one equation can be divided by the other to obtain:

$$R/R' = k/k \times [[H_2O_2]/[H_2O_2']]^n \times [[I^-]/[I^{-\prime}]]^m \text{ or}$$

$$\mathbf{R/R' = 1 \times [[H_2O_2]/[H_2O_2']]^n \times [[I^-]/[I^{-\prime}]]^m}$$

Log can be used to solve for m or n if either $[[H_2O_2]/[H_2O_2']]$ or $[[I^-]/[I^{-\prime}]]$ is **equal** to 1 (the concentrations were unchanged). For example if $[I^-]$ was the same in both runs, the log of both sides of the equation becomes: $\log (R/R') = \log 1 + n \log [[H_2O_2]/[H_2O_2']] + \log 1$. Only n would be **unknown**.

Experimental Rate-Law Expression: Rate of gas formation (mol/min) = k × [   ] [   ]
(The exponents should be left as fractional values. Show work below.)

Suggested Rate-Law Expression: Rate of gas formation (mol/min) = k × [   ] [   ]
(The exponents should be rounded to zero or whole number values. Show work below.)

POSTLAB QUESTIONS

1. Your suggested rate-law expression was: Rate of gas formation (mol/min) = k × [   ] [   ]
    (Use only zero or whole numbers for the exponents.)

2. If the vegetable-extract-catalyzed decomposition of hydrogen peroxide were to give the same rate-law expression, what would be the value of k for that reaction?

3. Compare the two catalysts. What can you conclude about the two different catalysts?

4. Suggest a mechanism that is consistent with the rate-law expression that you determined.

5. Draw a particle view of the reactants and products in the decomposition of hydrogen peroxide.

Date _____ **Student's Signature** _____

Kinetics of Decoloration of Crystal Violet

A Guided Inquiry Experiment

INTRODUCTION

During this experiment, you are to determine the numerical values of the rate-law expression for the decolorization of crystal violet. The methods to be used are different from the method used in most textbook problems, etc. Crystal violet is an organic dye that usually exists as a hybrid of two structures:

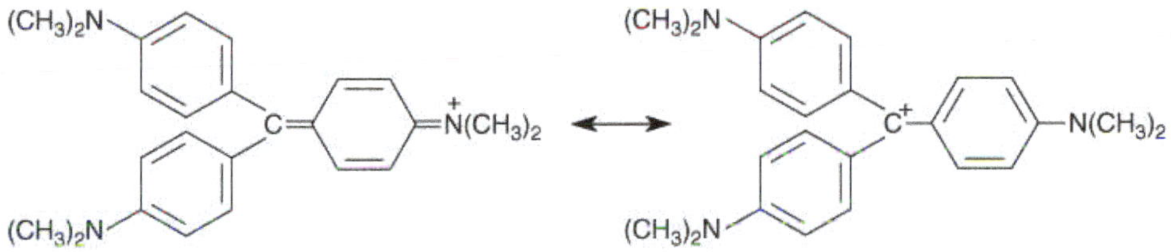

We will refer to the structure above as CV^+. CV^+ is highly colored. That means that it is a strong absorber of at least one region within the visible range of the electromagnetic spectrum. CV^+ reacts with hydroxide ion to produce the following compound referred to as CVOH:

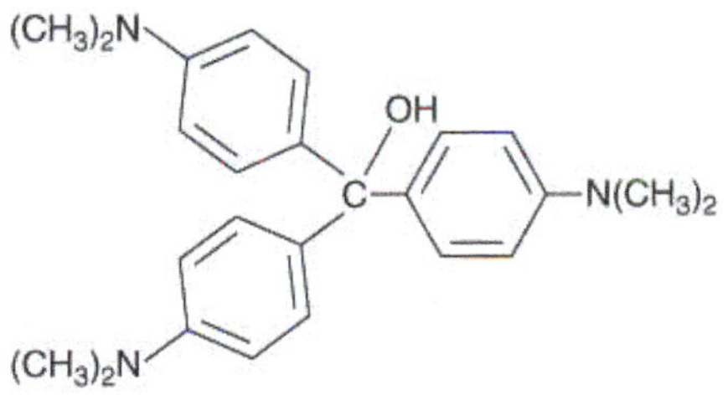

The reaction can be written as $CV^+ + OH^- \rightarrow CVOH$. CVOH is not an absorber in the visible range. Decolorization of the CV^+ can be used to follow the progress of the reaction. The rate-law expression for this equation is Rate $= k[CV^+]^m[OH^-]^n$. In this experiment, you are to determine the values of k, m, and n. Often, the values of m and n would be determined by studying the initial rates of reaction when only one concentration is varied from one experiment to the next. In those experiments, the concentration of each reactant is about the same concentration.

In this experiment, you are to run the experiment with the amount of hydroxide ion being three or more orders of magnitude above that which could possibly react. Being present in such a large excess, the $[OH^-]^n$ remains unchanged (a constant). If we combine k and $[OH^-]^n$, we have a new constant $k_{(observed)}$. The rate law now is Rate $= k_{(observed)}[CV^+]^m$.

The spectrophotometer measures absorbance or percent transmittance (%T) at various wavelengths within the visible spectrum (you may wish to review the section in the Common Procedures and Concepts at the end of this manual). In Experiment 17, students discovered that absorbance (A) is related to concentration via Beer's law: $A = \varepsilon bc$. $\varepsilon =$ molar absorptivity coefficient (a proportionality constant that is specific for the compound and wavelength used), b = length of the path through the sample (since you will be using the same or very similar cuvets throughout this experiment, b will be a constant and can be combined with the constant ε), and c = concentration. By measuring the absorbance of a solution of known concentration of crystal violet, one can obtain the value of εb. The units of εb will be 1/concentration. Once εb is known, it can be used to calculate concentration from absorbance values.

Integration of the rate-law expression of a **zero-order** reaction with a single concentration variable such as the expression

$$(\text{Rate} = -dc/dt = k_{(observed)}[CV^+]^0)$$

yields the integrated rate-law expression of:

$$[CV^+]_t = -k_{(observed)}t + [CV^+]_0.$$

A plot of $[CV^+]_t$ vs. t will yield a straight line with a slope of $-k_{(observed)}$.

Integration of the **first-order** rate-law expression

$$(\text{Rate} = -dc/dt = k_{(observed)}[CV^+]^1)$$

yields the integrated rate-law expression of:

$$\ln[CV^+]_t = -k_{(observed)}t + \ln[CV^+]_0.$$

A plot of $\ln[CV^+]_t$ vs. t will yield a straight line for a first-order reaction with the slope being $-k_{(observed)}$.

Integration of the **second-order** rate-law expression

$$(\text{Rate} = -dc/dt = k_{(observed)}[CV^+]^2)$$

yields the integrated rate-law expression of:

$$1/[CV^+]_t = k_{(observed)}t + (1/[CV^+]_0).$$

A plot of $1/[CV^+]_t$ vs. t will yield a straight line with a slope of $k_{(observed)}$.

Therefore, if one runs an experiment in which only one reactant has a variable concentration and if that reactant's concentration can be followed via measurements of absorbance, the concentration of the reactant and time may be plotted by the following three methods. The plot with the best straight line would indicate the order of the reaction. The following plots of a given data set indicate that the data is for a second-order reaction.

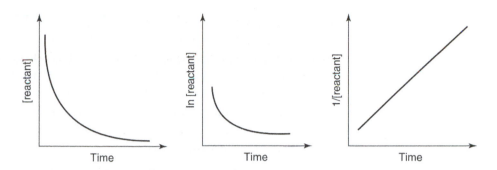

Having done the above, one would now know m in the equation: Rate = $k[CV^+]^m[OH^-]^n$. To determine n or the order with respect to hydroxide ion, one needs to recall $k_{(observed)} = k\ [OH^-]^n$. So the rate equation becomes: Rate = $k_{(observed)}\ [CV^+]^m$. If you have values for two runs in which $[OH^-]$ varies, the equation Rate = $k[CV^+]^m[OH^-]^n$ can be written twice and one can be divided by the other to yield:

$$\frac{Rate_2}{Rate_1} = \left(\frac{k[OH^-]_2}{k[OH^-]_1}\right)^n \left(\frac{[CV^+]_2}{[CV^+]_1}\right)^m$$

If k and $[CV^+]$ are not variables, both k/k and $[CV^+]_2/[CV^+]_1$ will equal 1. The log form of the equation is:

$$\log\left(\frac{Rate_2}{Rate_1}\right) = n\ \log\left(\frac{[OH^-]_2}{[OH^-]_1}\right)$$

If the Rate and $[OH^-]$ values are known for the two experiments, you can solve for n. Then k can be solved by using $k_{(observed)} = k\ [OH^-]^n$ for one of the experiments.

OBJECTIVES

Knowing the information in the introduction above, your task is to design experiments so that you will be able to construct an absorbance curve for crystal violet (like the one given for Cu^{2+} in the Common Procedures and Concepts Section at the end of this manual), select the wavelength at which crystal violet has maximum absorbance, and run three experiments in which you measure absorbance versus time. From these experiments, you are to determine the rate-law expression for the decolorization of crystal violet, its rate constant, and the order of each reactant.

CONCEPTS AND TECHNIQUES

This experiment uses stoichiometric concepts, solution preparation, spectrometry, and kinetics.

ACTIVITIES

You will prepare crystal violet solutions in 50% ethanol that can be used in the spectrometer without their absorbance being off the scale or too low to read. You will need to make a number of spectrophotometric determinations, use Beer's law to determine concentration, prepare several plots relating concentration versus time, and use your results to determine the numerical values in the rate-law expression.

> **CAUTION**
>
> **You will be handling crystal violet solutions. Crystal violet is an organic dye and it will stain most materials that it contacts. Wear eye protection at all times.**

PROCEDURES

Preparing Crystal Violet Solution

1-1. Find the transmittance and absorbance of the 8.0×10^{-5} M CV^+ stock solution at 500 nm. Refer to the section at the end of this manual for the correct use of the spectrophotometer. You will need one cuvet with solvent to be used when adjusting the 100% transmittance and another cuvet for the solution you are testing. The cuvet with the solvent will be kept to use throughout the experiment. Find εb at 500 nm and record this value.

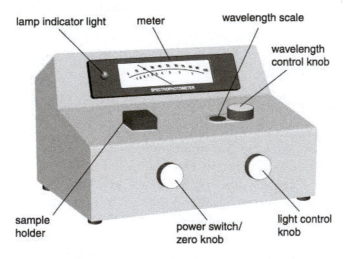

> **NOTE:** Remember to match the mark on the cuvet with the mark at the top of the sample holder and to close the sample holder cover.

1-2. From your findings above, calculate the concentration of CV^+ needed to yield an absorbance of 0.4 at a wavelength of 500 nm. Prepare 30 mL of this solution to use throughout the experiment from the 8.0×10^{-5} M CV^+ stock solution. Record the actual labeled concentration of the stock solution provided.

Absorbance Curve

2-1. Using the sample you just prepared, determine the solution's absorbance every 20 nm from 400 nm to 700 nm.

2-2. For your Report Form you will need to carefully plot absorbance vs. wavelength to obtain a plot like that found in the Common Procedures and Concepts Section at the end of this manual. From such a plot you could carefully determine the wavelength at which maximum absorbance occurs. Record the wavelength as the wavelength at which you will run the remaining analyses. Calculate the ε at this wavelength (b = 1 cm). Calculate the concentration of crystal violet that you would expect to yield an absorbance between 0.8 and 1.0 at that wavelength.

Decolorization

3-1. Make the solutions for Procedures 3-1 through 3-4 from the crystal violet solution you made in Procedure 1-2, the stock NaOH solution, and the 50% ethanol solution. Record the concentration of the stock NaOH solution. Prepare the solutions needed in which the concentrations of crystal violet will be about that which you determined in Procedure 2-2 and sodium hydroxide in 50% ethanol will be 0.10 *M* after mixing to produce a total volume of 7.0 mL. In a test tube, mix the solutions and observe any reaction that takes place. Record your observations.

3-2. Using the solutions, spectrophotometer, cuvets, etc., needed to run a determination in which the concentrations of crystal violet will be about that which you determined in Procedure 2-2 and sodium hydroxide in 50% ethanol will be 0.10 *M* after mixing, prepare for the actual run using the spectrophotometer. Organize the next few steps so that in the least amount of time you can mix the solutions, load the cuvet, and place the cuvet into a spectrophotometer that already has the wavelength set and the 0 and 100% transmittance adjusted. Start timing your reaction as soon as you mix the reagents. Record % transmittance values every 15 seconds for 5 minutes or until the % transmittance becomes too small to clearly distinguish one from another. Remember the volume of the cuvet is 7 mL. This is Run #1.

3-3. Prepare the solutions, spectrophotometer, cuvets, etc., needed to run a second determination in which the initial concentration of crystal violet will be about that which you determined above and the concentration of sodium hydroxide in 50% ethanol has been adjusted with the goal of having the decolorization take place over a period of 4 or 5 minutes. Again, organize your next few steps so that in the least amount of time you can mix the solutions, load the cuvet, and place the cuvet into a spectrophotometer that already has the wavelength set and the 0 and 100% transmittance adjusted. Start timing your reaction as soon as you mix the reagents. Record % transmittance values every 15 seconds for 5 minutes or until the change in % transmittance becomes too small to clearly distinguish one from another. Remember the volume of the cuvet is 7 mL. This is Run #2.

3-4. Repeat Procedure 3-2 but adjust the concentration of NaOH so that the reaction time will differ by about 30%. This is Run #3.

3-5. Have your instructor sign your notebook and Report Form.

Name (Print)   Date (of Lab Meeting)   Instructor
_____   _____ _____

Course/Section
_____

21   EXPERIMENT 21: KINETICS OF DECOLORATION OF CRYSTAL VIOLET

Prelab Exercises

1. What is a cuvet? Limit your answer to a definition that applies to this experiment.

2. Why is there a vertical line on the side of the cuvet?

3. State Beer's law and define each of the terms used.

4. If a cuvet holds 7 mL, how would you prepare 7.0 mL of solution that is 1.5×10^{-5} M CV^+ and 0.1 M NaOH in 50% ethanol from 8.0×10^{-5} M CV^+ in 50% ethanol, 1.0 M NaOH in 50% ethanol, and 50% ethanol?

5. If the absorbance of a solution is 0.79 at 480 nm and it is 3.5×10^{-7} M, what is the value and units of εb at the wavelength at which the absorbance was read?

Date _____ **Student's Signature** _____

21 EXPERIMENT 21: KINECTICS OF DECOLORATION OF CRYSTAL VIOLET

Report Form

DATA

Solution of crystal violet having an absorbance of about 0.4 at a wavelength of 500 nm

Ɛb at 500 nm: _____

Concentration of stock CV^+ solution: _____ M

Volume of stock CV^+ solution used: _____ mL

Volume of stock 50% ethanol solution used: _____ mL

Concentration of crystal violet solution prepared: _____ M

Absorbance Curve

Wavelength/%Transmittance:

400 nm	%T =	480 nm	%T =	560 nm	%T =	640 nm	%T =
420 nm	%T =	500 nm	%T =	580 nm	%T =	660 nm	%T =
440 nm	%T =	520 nm	%T =	600 nm	%T =	680 nm	%T =
460 nm	%T =	540 nm	%T =	620 nm	%T =	700 nm	%T =

Wavelength selected for remaining determinations: _____ nm

Ɛb at this wavelength: _____

Decolorization

Concentration of stock NaOH solution: _____ M

Observations in Procedure 3-1:

Run #1

Vol. of prepared crystal violet solution used _____ mL

Vol. of prepared NaOH solution used _____ mL

Vol. of 50% ethanol solution used _____ mL

% Transmittance/Time:

sec	%T =	sec	%T =	sec	%T =	sec	%T =
sec	%T =	sec	%T =	sec	%T =	sec	%T =
sec	%T =	sec	%T =	sec	%T =	sec	%T =
sec	%T =	sec	%T =	sec	%T =	sec	%T =
sec	%T =	sec	%T =	sec	%T =	sec	%T =
sec	%T =	sec	%T =	sec	%T =	sec	%T =

Run #2

Vol. of prepared crystal violet solution used _____ mL

Vol. of prepared NaOH solution used _____ mL

Vol. of 50% ethanol solution used _____ mL

% Transmittance/Time:

sec	%T =	sec	%T =	sec	%T =	sec	%T =
sec	%T =	sec	%T =	sec	%T =	sec	%T =
sec	%T =	sec	%T =	sec	%T =	sec	%T =
sec	%T =	sec	%T =	sec	%T =	sec	%T =
sec	%T =	sec	%T =	sec	%T =	sec	%T =
sec	%T =	sec	%T =	sec	%T =	sec	%T =

Run #3

Vol. of prepared crystal violet solution used _____ mL

Vol. of prepared NaOH solution used _____ mL

Vol. of 50% ethanol solution used _____ mL

% Transmittance/Time:

sec	%T =		sec	%T =		sec	%T =		sec	%T =
sec	%T =		sec	%T =		sec	%T =		sec	%T =
sec	%T =		sec	%T =		sec	%T =		sec	%T =
sec	%T =		sec	%T =		sec	%T =		sec	%T =
sec	%T =		sec	%T =		sec	%T =		sec	%T =
sec	%T =		sec	%T =		sec	%T =		sec	%T =

Date _____   **Instructor's Signature** _____

ANALYSIS

Absorbance Curve

Attach tables for Absorbance Curve and Runs #1, #2, and #3 in which your data has been converted from % transmittance to absorbance.

Of the wavelengths considered in the table (400 nm to 700 nm in 20 nm increments), which yielded

the maximum absorbance? _____ nm

Attach a plot of absorbance versus wavelength. Absorbance is to be the vertical axis and wavelength the horizontal axis.

From your graph, what wavelength yielded the maximum absorbance? _____
(Indicate this on your graph.)

Use Beer's law and calculate the value of εb at the above wavelength. _____
(Show your work.)

Run #1

Vol. of prepared crystal violet solution used   _____ mL

Vol. of prepared NaOH solution used   _____ mL

Vol. of 50% ethanol solution used   _____ mL

Concentration of crystal violet after mixing but before any has decolorized: _____ *M*
(Show work.)

Concentration of NaOH in the solution after mixing but before any has reacted: _____ *M*
(Show work.)

(Use Beer's law ($A = \varepsilon bc$) and the value of εb calculated above to calculate the concentration of crystal violet at various times during this decolorization reaction.)

sec	c =	sec	c =	sec	c =	sec	c =
sec	c =	sec	c =	sec	c =	sec	c =
sec	c =	sec	c =	sec	c =	sec	c =
sec	c =	sec	c =	sec	c =	sec	c =
sec	c =	sec	c =	sec	c =	sec	c =
sec	c =	sec	c =	sec	c =	sec	c =

For one data point, show your calculation of concentration from the absorbance value on a separate sheet.

Run #2

Vol. of prepared crystal violet solution used   _____ mL

Vol. of prepared NaOH solution used   _____ mL

Vol. of 50% ethanol solution used   _____ mL

Concentration of crystal violet after mixing but before any has decolorized: _____ *M*
(Show work.)

Concentration of NaOH in the solution after mixing but before any has reacted: _____ *M*
(Show work.)

(Use Beer's law ($A = \varepsilon bc$) and the value of εb calculated above to calculate the concentration of crystal violet at various times during this decolorization reaction.)

sec	c =	sec	c =	sec	c =	sec	c =
sec	c =	sec	c =	sec	c =	sec	c =
sec	c =	sec	c =	sec	c =	sec	c =
sec	c =	sec	c =	sec	c =	sec	c =
sec	c =	sec	c =	sec	c =	sec	c =
sec	c =	sec	c =	sec	c =	sec	c =

Run #3

Vol. of prepared crystal violet solution used _____ mL

Vol. of prepared NaOH solution used _____ mL

Vol. of 50% ethanol solution used _____ mL

Concentration of crystal violet after mixing but before any has decolorized: _____ *M*
(Show work.)

Concentration of NaOH in the solution after mixing but before any has reacted: _____ *M*
(Show work.)

(Use Beer's law ($A = \varepsilon bc$) and the value of εb calculated above to calculate the concentration of crystal violet at various times during this decolorization reaction.)

sec	c =	sec	c =	sec	c =	sec	c =
sec	c =	sec	c =	sec	c =	sec	c =
sec	c =	sec	c =	sec	c =	sec	c =
sec	c =	sec	c =	sec	c =	sec	c =
sec	c =	sec	c =	sec	c =	sec	c =
sec	c =	sec	c =	sec	c =	sec	c =

Plots

Attach plots of the first 75 seconds of Run #1 in which you have plotted c vs. time, ln(c) vs. time, and 1/c vs. time as described in the Introduction. Select the plot that is closest to being a straight line. This will indicate the value of the exponent of the crystal violet's concentration term in the rate-law expression. The exponent of the crystal violet term in the rate-law expression is: _____.

Attach plots of the first 75 seconds of Runs #2 and #3 graphed as c vs. time. Measure the initial rate (slope) of each. Use the slopes as the rates of reaction to solve for the exponent of hydroxide ion:

log (rate$_3$/rate$_2$) = n log ((initial concentration of NaOH)$_3$/(initial concentration of NaOH)$_2$).

	[CV$^+$]	[OH$^-$]	Initial Rate (instantaneous rate at t=0)
Experiment 2	same		
Experiment 3	same		

The exponent of the hydroxide ion is _____. Show your calculation below.

You now have the exponents on both the crystal violet and hydroxide. Solve for the value of the rate-law expression constant, k, using Experiments 2 or 3.

The value for this constant is _____. Show your calculation below.

You should now be able to write the general rate-law expression for the decolorization of crystal violet in a sodium hydroxide solution with specific values for all terms except the rate and the concentrations. Give this below.

POSTLAB QUESTIONS

1. Your suggested rate-law expression using only zero or whole numbers for the exponents is:

2. What is the overall order of this reaction?

3. What does the rate law in Postlab Question 1 tell you about the relationship between the rate and the concentrations?

4. Assume you are helping a fellow student. Explain to "Chris" how to determine a rate-law expression that includes values for the exponents and constant.

5. Draw a particle view of the decolorization of crystal violet in a sodium hydroxide solution.

Date _____ **Student's Signature** _____

Factors Affecting Reactions

A Guided Inquiry Experiment

INTRODUCTION

In this course, you have made calculations based upon a number of chemical reactions. You have determined which reactants were the limiting reactants, the percent yield, etc. For most of those reactions, you considered the reactions as going to completion and that only the products remained after the reaction had run to completion. But, is this what actually happened? In this experiment, we will investigate the reactants and products throughout a chemical reaction.

OBJECTIVES

In this experiment, you will investigate the relationship between the reactants and products in a reaction. You will study the effects of temperature and the addition of other substances.

CONCEPTS

This experiment uses the concepts of net ionic equations, detection of limiting reactants, tests for completeness of reaction, and influence of competing reactions.

TECHNIQUES

Hot water and ice water baths are some of the techniques that you will use during this experiment. You will use visual observations to determine completeness of reactions. Small quantities (microscale or small-scale) techniques will be use throughout this experiment.

ACTIVITIES

In this experiment, you will investigate the relationship between the reactants and products of a reaction. Then you will investigate the relationship between temperature and the products of a reaction. Based upon your observations, you will make conclusions about completeness of the reaction studied, reversibility of the reactions, and possible competing reactions.

CAUTION

Use approved eye protection. Be aware of chemicals and hot equipment.

PROCEDURES

Observing a Reaction

1-1. Measure 10 mL of distilled water with a 10-mL graduated cylinder.

1-2. Obtain a dropper bottle of 1 M NH4SCN (in 0.3 M HNO$_3$) and a dropper bottle of 1 M Fe(NO$_3$)$_3$ (in 0.3 M HNO$_3$). Note the color of each solution.

1-3. Add 1 drop of the NH$_4$SCN and 1 drop of the Fe(NO$_3$)$_3$ to the water in the graduated cylinder. Record your observations. Do you see any evidence that a chemical reaction has occurred? You should stir the solution with a clean transfer pipet.

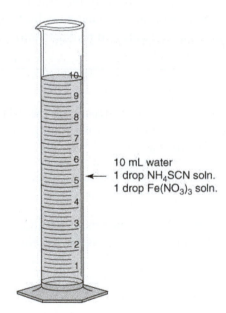

10 mL water
1 drop NH$_4$SCN soln.
1 drop Fe(NO$_3$)$_3$ soln.

1-4. The net ionic equation for the reaction you observed is $SCN^- + Fe^{3+} \rightarrow$ FeSCN^{2+}. Identify the color of each substance in the reaction. Save the graduated cylinder with its contents until the conclusion of the experiment.

Effects of Temperature

2-1. Fill three small test tubes ⅓ full of the solution from Procedure 1-3. Retain the remaining solution in the graduated cylinder. Compare the color of the three test tubes.

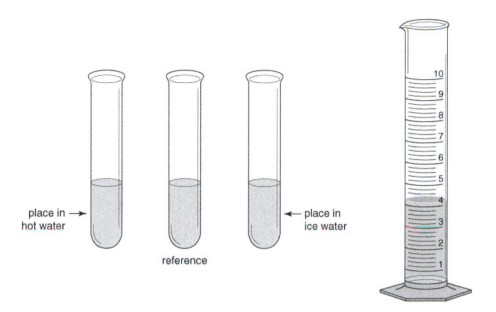

place in → hot water

place in ← ice water

reference

2-2. Prepare a hot water bath in a beaker in which your small test tube can stand upright. Also prepare an ice water bath in a similar sized beaker. Place one test tube into an ice water bath, one into a hot water bath, and leave one test tube at room temperature. Let these tubes remain undisturbed for about 10 minutes. Proceed with the next section. You will be reminded to check these later.

Investigating the Reaction

3-1. Using the transfer pipet, put five drops of the solution from Procedure 1-3 into six different spots of a spot plate (a well plate can also be used). Retain the remaining solution.

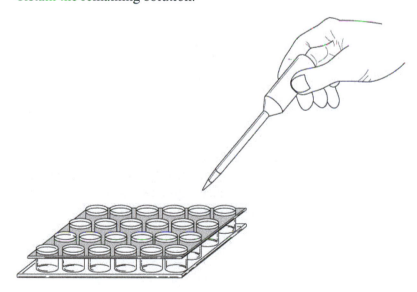

3-2. To the solution in the first spot, add one drop of the 1 *M* SCN⁻ solution. Record your observations. Stir with a clean toothpick. Compare this spot's solution to those of the other spots. Note any color changes.

3-3.   Into the third spot, add one drop of the 1 M Fe^{3+} solution. Record your observations. Stir with a clean toothpick. Compare the color of the solutions in the second and third spots.

3-4.   You will continue to use the second spot as a reference of the original color of your solution.

Effects of Other Reagents

4-1.   To the fourth spot you prepared in Procedure 3-1, add a drop of the 1 M $NaNO_3$ solution. Stir with a clean toothpick. Record your observations. Compare the color of this spot to the second spot that you are using for a reference. Do you see evidence of further reaction or just the results of dilution?

4-2.   To the fifth spot, add a drop of the 1 M Na_2HPO_4 solution. Stir with a clean toothpick. Record your observations. Compare the color of this spot to the reference spot. Do you see evidence of further reaction or just the results of dilution?

4-3.   To the sixth spot, add a drop of the 1 M Na_2HPO_4 solution as you did in Procedure 4-2. Stir with a clean toothpick.

4-4.   Once you have two identical solutions in the fifth and sixth spots, add a drop of the 1 M SCN^- solution to the fifth spot. Record your results. To the sixth spot, add a drop of the 1 M Fe^{3+} solution. Record your results and observations of the colors produced.

4-5.   Dispose the contents as your instructor directs. Clean and dry the spot plate and all equipment and glassware.

Return to Effects of Temperature

5-1.   About 10 minutes or more after completion of Procedure 2-2, record your observations. Compare the color of the hot and cold tubes with the room-temperature tube.

5-2.   Clean the hot water and ice water beakers and test tubes. Dispose of the contents as your instructor directs. Have your instructor sign your notebook and Report Form.

22 EXPERIMENT 22: FACTORS AFFECTING REACTIONS

Prelab Exercises

1. What is a net ionic equation?

2. Explain a "limiting reactant" and how it differs from the other reactants.

3. Explain the meaning of "the reaction goes to less than 100% completion."

4. For the each of the following reactant pairs, give the balanced molecular equation, then give the net ionic equation.

   a. ___$BaBr_2(aq) +$ ___$K_2CO_3(aq) \rightarrow$

   b. ___$NaOH(aq) +$ ___$H_2SO_4(aq) \rightarrow$

5. Answer any question added by your instructor.

Date _____   **Student's Signature** _____

Name (Print)   Date (of Lab Meeting)   Instructor

Course/Section   Partner's Name (If Applicable)

22   EXPERIMENT 22: FACTORS AFFECTING REACTIONS

Report Form

DATA

Observing a Reaction

Give your observations for each below:

  Stock 1 M NH$_4$SCN solution:

  Stock 1 M Fe(NO$_3$)$_3$ solution:

  Identify the color of each substance in the reaction.
    SCN$^-$, Fe^{3+}, FeSCN^{2+}

Investigating the Reaction

Give your observations when additional SCN$^-$ was added:

Give your observations when additional Fe^{3+} was added:

Effects of Temperature

Give your observations for each below:

hot water bath

ice water bath

Effects of Other Reagents

Give your observations for each below:

Addition of 1 M NaNO$_3$

Addition of 1 M Na$_2$HPO$_4$

Addition of 1 M Na$_2$HPO$_4$ followed by the addition of SCN$^-$

Addition of 1 M Na$_2$HPO$_4$ followed by the addition of Fe^{3+}

Date _____ **Instructor's Signature** _____

ANALYSIS

Observing a Reaction

Describe the evidence you have that this chemical reaction took place.

$$SCN^- + Fe^{3+} \rightarrow FeSCN^{2+}$$

Draw a particle view of the reaction between $SCN^- + Fe^{3+}$.

Investigating the Reaction

When additional SCN^- was added, what ion had to be present in the solution to account for your observations?

When additional Fe^{3+} was added, what ion had to be present in the solution to account for your observations?

Before the addition of extra reactants, was the reaction complete? Explain why or why not.

Before you added more drops of each of the reactants, what substances were in the test tube? Explain your reasoning.

Generally, what was the effect of adding more of each reactant?

Effects of Temperature

What was the effect of increased temperature? Were more or less products produced in the hot solution? Explain your reasoning.

What was the effect of decreased temperature? Were more or less products obtained in the cold solution? Explain your reasoning.

When a reaction is reversible, the products become reactants and the reactants become products. If the reaction is less than 100% complete, an equilibrium may exist. We show an equilibrium with double arrows:

$$SCN^- + Fe^{3+} \Leftrightarrow FeSCN^{2+}$$

Do your observations for the addition of extra reactants and temperature change indicate that this reaction is in equilibrium? Explain.

What was the effect of the addition of more reactant?

If heat were included in the equation, would you place it as a product or a reactant (endothermic or exothermic)? Explain your reasoning.

Effects of Other Reagents

What was the effect of adding 1 M $NaNO_3$? Were more or less products present in the solution after this addition? Explain your reasoning.

What was the effect of adding 1 M Na_2HPO_4? Were more or less products present in the solution after this addition? Explain your reasoning.

Based on your observations of the addition of 1 M Na_2HPO_4 followed by the addition of SCN^- and your observations of the addition of 1 M Na_2HPO_4 followed by the addition of Fe^{3+}, what would you say the HPO_4^{2-} reacted with?

The PO_4^{3-} ion is known to complex with metal ions. Do your observations support this property of phosphate? Explain.

POSTLAB QUESTIONS

1. When a reaction produces more products, we say that it shifts to the right. If the reaction exists as an equilibrium, which changes caused a shift to the right (addition of a reactant, increased heat, decreased heat, addition of $NaNO_3$, or addition of Na_2HPO_4)?

2. When a reaction produces less products, we say that it shifts to the left. If the reaction exists as an equilibrium, which changes caused a shift to the left (addition of a reactant, increased heat, decreased heat, addition of $NaNO_3$, or addition of Na_2HPO_4)?

3. The HSCN is a weak acid. What effect would the addition of a strong acid have on the system studied in this experiment?

4. The ammonium and nitrate ions do not participate in the reaction you studied. What term(s) describe(s) ions that have this role?

5. What is meant by chemical equilibrium?

6. What is meant by Le Chatelier's principle? See the definition in your textbook.

Date _____ Student's Signature _____

Formation of a Complex

A Skill Building Experiment

INTRODUCTION

When an electron pair of an electron-rich nonmetal group is shared with a metal atom or ion, a coordination compound or complex is formed. The nonmetal group is referred to as a ligand. When complexes are prepared, their composition may be unknown. If the complex can be isolated and purified, its composition can be determined via a number of methods. Elemental analyses performed by commercial laboratories; magnetic measurements; information from ultraviolet, visible, and infrared spectra; data obtained by nuclear magnetic resonance spectroscopy; and single-crystal X-ray crystallography may provide information about the chemical nature and valence state of the central metal atom, the type of ligands, and their mode of bonding. However, a large number of complexes, including many that exist in biological systems, cannot be isolated. They exist only in solution and attempting to isolate them often changes their composition. One such complex that is used by analytical chemists to detect the presence of dissolved iron ion is the deep red coordination compound formed when Fe^{3+} and SCN^- (thiocyanate) ions react. Fortunately, techniques exist that allow us to determine the composition of such non-isolatable complexes. Job's method of continuous variation (the method used in this experiment) can be used to study non-isolatable complexes.

The spectrophotometer measures absorbance or % transmittance at various wavelengths within the visible spectrum. (You may wish to review Experiment 17, Experiment 21, and the Common Procedures and Concepts Section at the end of this manual.) Absorbance (A) is related to concentration via Beer's law: $A = \varepsilon bc$. ε = molar absorptivity coefficient (a proportionality constant that is specific for the compound and wavelength used). b = length of the path through the sample. (Since you will be using the same or very similar cuvets throughout this experiment, b will be a constant and can be combined with the constant ε.) c = concentration. By measuring the absorbance of a solution of known concentration of complex, one can obtain the value of εb. The units of εb will be 1/concentration. Once εb is known, it can be used to calculate concentration from absorbance values for other samples.

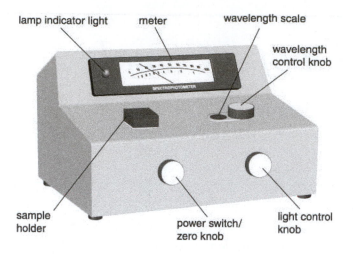

lamp indicator light   meter   wavelength scale

wavelength control knob

sample holder

power switch/ zero knob

light control knob

OBJECTIVES

This experiment provides the opportunity to become acquainted with a method of determining the composition of a complex without having to isolate the complex. The method is known as Job's method of continuous variation.

CONCEPTS

This experiment uses the concepts of solution concentration, spectroscopy, and reactions. The spectrometer is used throughout this experiment.

TECHNIQUES

Diluting a solution, reading a spectrophotometer, calculating mole fractions, and graphing experimental data are some of the techniques that you will use during this experiment. (See the Common Procedures and Concepts Section at the end of this manual.)

ACTIVITIES

In this experiment, you will collect spectrophotometric data related to reactions of known amounts of Fe^{3+} and SCN^-. From relationships between the mole fractions of each, the formula of the complex can be determined.

CAUTION

Use approved eye protection. Practice proper laboratory techniques.
Keep a clean work area. Dispose of chemicals as directed by your instructor.

PROCEDURES

1-1. Obtain two burets. Rinse the burets with distilled water and allow them to drain through the stopcock. Examine the inside of the burets. If there are water drops clinging to the inside wall, clean the buret. Label one buret "Fe"; label the other "SCN."

1-2. Take two appropriately labeled, clean and dry flasks or beakers to your instructor to obtain 30 mL each of 0.0016 M Fe(NO)$_3$ in 0.1 M HNO$_3$ and 0.0016 M KSCN in 0.1 M HNO$_3$ solutions. (The concentration of the solutions should be known to three significant digits.) Record the concentration of each solution in your notebook. Rinse and fill each buret about half full with its designated solution. Discharge any air bubbles that might be trapped in the buret tip.

1-3. Obtain nine small, clean, and dry matching test tubes with a volume of 8 to 10 mL. Label the tubes "1," "2," ... "9."

1-4. Add to each test tube the volume of Fe^{3+} and SCN$^-$ solutions given in the following table. Record the exact volumes used in your notebook. Thoroughly mix the contents of each tube by rocking the tube and vigorously tapping it several times. *Mixing is important!*

Solution (Tube #)

Regent	1	2	3	4	5	6	7	8	9
mL Fe^{3+}	0	1	2	2.5	3	3.5	4	5	6
mL SCN$^-$	6	5	4	3.5	3	2.5	2	1	0
Total mL	6	6	6	6	6	6	6	6	6

1-5. Obtain two cuvets. Turn on the spectrophotometer, allow it to warm up, and set the wavelength to 450 nm. With the empty sample chamber closed, adjust the "zero transmittance" knob until the dial reads zero transmittance. Place the contents of test tube #1 into the cuvet and place it in the sample chamber. Adjust the "100% transmittance" knob until the dial reads 100% transmittance (absorbance = 0). Remove the cuvet with its "tube #1" contents and adjust the "zero transmittance" knob until the dial reads zero transmittance. Place the cuvet with its "tube #1" contents back into the sample chamber. Adjust the "100% transmittance" knob until the dial reads 100% transmittance (absorbance = 0).

Repeat the zero-transmittance and 100% transmittance adjustments until the dial remains at zero (chamber closed, empty) and at 100 (cuvet with the contents of test tube #1 in the sample holder chamber). **Then do not touch these knobs again until you have finished the experiment.**

NOTE: Do not touch the middle sections of the cuvets with your fingers. Grease or oil on the surface will affect the absorbance measurements.

1-6. One test tube at a time, transfer the contents of test tubes #2 through #9 to the second cuvet and measure their % transmittance. Record all % transmittance values in your notebook. Have your instructor sign your notebook and Report Form. For your Report Form, you will need to convert % transmittance values to absorbance values (see the Common Procedures and Concepts Section at the end of this manual for the relationship between % transmittance and absorbance). In the Report Form, you will use your data to construct a plot similar to the one below. This type of plot is often called a Job's plot.

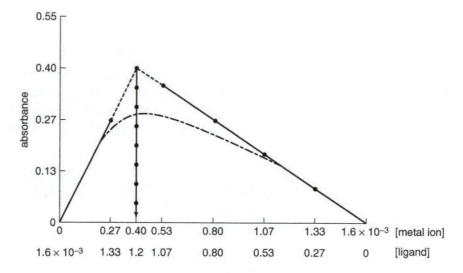

This is an example of a Job's plot of a metal ion and a ligand that forms a complex.

Name (Print) _____ Date (of Lab Meeting) _____ Instructor _____

Course/Section _____

23 EXPERIMENT 23: FORMATION OF A COMPLEX

Prelab Exercises

1. Define:

   a. spectrophotometer

   b. absorbance

   c. cuvet

   d. coordination compound

   e. ligand

2. Explain how the absorbance and transmittance measurements will be affected if there was grease residue in the middle of the cuvet. Give your rationale. How would this compare with touching the middle of the cuvet?

3. State Beer's law and define each of the terms.

Date _____ **Student's Signature** _____

23   **EXPERIMENT 23: FORMATION OF A COMPLEX**

Report Form

DATA

Concentration of the Fe^{3+} solution (3 significant figures): _____ *M*

Concentration of the SCN^- solution (3 significant figures): _____ *M*

Exact volumes of stock solutions used and % transmittance values obtained

Tube #	1	2	3	4	5	6	7	8	9
mL Fe^{3+}	0.00								
mL SCN^-									0.00
% transmit									

Date _____   **Instructor's Signature** _____

ANALYSIS

Complete the following table. You will need to convert % transmittance to absorbance and to calculate the initial concentrations of the ions ($[Fe^{3+}]_i$ and $[SCN^-]_i$). These are the concentrations after mixing but before any complex is formed.

Tube #	1	2	3	4	5	6	7	8	9
Absorbance									
$[Fe^{3+}]_i$	0.00								
$[SCN^-]_i$									0.00

What is the relationship between the values for $[Fe^{3+}]_i$ and $[SCN^-]_i$?
(HINT: Look at ($[Fe^{3+}]_i \times [SCN^-]_i$), ($[Fe^{3+}]_i/[SCN^-]_i$), ($[Fe^{3+}]_i + [SCN^-]_i$), and $[Fe^{3+}]_i - [SCN^-]_i$).

Explain why this relationship is important.

Attach a plot of absorbance versus $[Fe^{3+}]_i$, similar to the sketch in Procedure 1-6. Connect the points at the two sides of the plot by separate straight lines (use only the values for tubes #1 to #3 and #7 to #9 for this purpose). Extend the two lines until they meet somewhere between the two extremes of $[Fe^{3+}]_i$. What is the $[Fe^{3+}]_i$ at which the two lines meet?

Calculate the concentration of $[SCN^-]_i$ corresponding to the $[Fe^{3+}]_i$ at which the two lines meet.

From the ratio of the concentrations of $[SCN^-]_i$ and $[Fe^{3+}]_i$ that yield the maximum concentration of complex and therefore the maximum absorbance, one can determine the ratio of ligand to metal ion. You may have to round up or round down to obtain a ratio that is small whole numbers. Your data indicates that the ratio is (as small whole numbers) _____.

Based upon the plot of absorbance versus ($[Fe^{3+}]_i$ or $[SCN^-]_i$) the formula and charge of the complex is:

The equation for the formation of the product is:

Since there was much more thiocyanate ion than iron ion present in tube #2, one can assume that the concentration of complex formed would equal the concentration of the iron initially present. This concentration, Beer's law, and the absorbance for tube #2 can be used to determine the value of εb. The units of εb will be 1/concentration. Show calculation of εb.

Once εb is known, it can be used to calculate concentration of the complex from absorbance values. Once the concentration of complex is known, you should be able to calculate the concentrations of the reactants that remain when the system is at equilibrium. Complete the following table:

Tube #	1	2	3	4	5	6	7	8	9
Absorbance									
[complex]$_{eq}$	0.00								0.00
[Fe^{3+}]$_{eq}$	0.00								
[SCN$^-$]$_{eq}$									0.00
εb	XX		XX	XX	XX	XX	XX	XX	XX

For tubes 3, 5, and 8, calculate the equilibrium constant for the formation of the complex (K_f) _____ _____ _____ (Show your calculations.) HINT: Before you can write an equilibrium expression, you must use a balanced chemical equation. Experimentally obtained constants should vary but still have about the same order of magnitude.

POSTLAB QUESTIONS

1.

    a. From your experimental results, calculate the difference between the "rounded ratio" of [SCN$^-$]$_i$/[Fe^{3+}]$_i$, and "experimental ratio." What is this difference as a percentage of the "rounded ratio"?

b. Calculate the "theoretical ratios of initial concentrations" for reactants in the formation of $[Fe(SCN)_4]^-$.

c. Assuming that the difference "rounded" to "experimental ratio" may be as large as 10%, can Job's method distinguish the complexes with $n = 6$ and 7? Explain your answer.

2. Draw a Lewis formula of the complex having the formula you determined in this experiment.

3. Draw particle views of the reactants and products of the formation equation for the formation of the complex whose formula you determined.

4. Based on the value of the K_f, describe on the molecular level what would happen if more SCN ions were added.

Date _____ **Student's Signature** _____

Acids and pH

A Guided Inquiry Experiment

INTRODUCTION

The measurement of the acid–base content of a substance can be accomplished a number of ways: litmus paper, pH paper, indicators, or with a pH meter. An acid solution has a pH below 7, while a basic solution has a pH above 7, and a neutral solution has a pH of 7. The pH of a solution is related to the amount of hydrogen ion by: $pH = -\log [H^+]$.

OBJECTIVES

In this experiment, you will investigate the relationship between various methods of determining acid–base characteristics. Then, you will investigate pH changes with concentration and with the addition of a salt.

CONCEPTS

This experiment uses the concepts of solution concentration, molarity, and ionization.

TECHNIQUES

Diluting a solution, calibrating and reading a pH meter, and serial dilution are some of the techniques that you will use during this experiment.

ACTIVITIES

In this experiment, you will find the pH of various substance via the pH meter, plus their reaction with litmus paper and with two other indicators. Then, you will investigate the relationship between pH and concentration of acetic acid. Finally, you will investigate the changes in pH due to addition of a salt.

CAUTION

Use approved eye protection. Be aware of chemicals and hot equipment.

PROCEDURES

Methods of Collecting Acid–Base Information

1-1. Obtain a spot plate or a well plate and dropper bottles of 0.10 M solutions of HCl, HNO$_3$, NaOH, KHP, H$_2$C$_2$O$_4$, HC$_2$H$_3$O$_2$, KOH, distilled water, Ba(OH)$_2$, and H$_2$SO$_4$. If dropper bottles of these solutions are not available, obtain about 3 mL of each solution in small beakers or test tubes.

1-2. Place the plate on a sheet of paper. You can make notes about the contents of each well on this sheet. Fill two spots about ½ full of the first solution.

1-3. Test the first spot of each with a small portion of a strip of red litmus paper. Repeat with blue litmus paper. Add a drop of phenolphthalein indicator to the solution in the spot. Record all results.

1-4. Calibrate the pH meter assigned to you using the buffer solution(s) provided. Follow the instructions provided for the type of pH meter in use in your laboratory. See the Common Procedures and Concepts Section at the end of this manual. If a pH meter is not available, your instructor may ask you to use pH paper.

1-5. Measure the pH of the second spot. Then add 1 drop of bromothymol blue indicator solution to the second spot. Record both of your results.

1-6. Repeat Procedures 1-2 through 1-5 with each of the remaining solutions. Depending on your equipment, you may not need to calibrate the pH meter for each solution. Empty and clean your spot plate as needed. Use a 600-ml beaker for waste.

1-7. When you have finished, empty the spot or well plate and beakers or test tubes as directed by your instructor. Rinse the plate and invert it. Clean any beakers or test tubes you may have used.

pH and Concentration

2-1. Obtain a small beaker with about 20 mL of 0.100 M acetic acid. Measure the pH of this solution. Since this solution was prepared for you, the beaker can be shared by a number of groups.

2-2. Obtain a small beaker and two droppers. You will use one dropper for water and one dropper for acetic acid; remember to rinse it thoroughly when changing molarities.

2-3. You will need to prepare 0.010 M, 0.0010 M, and 0.00010 M solutions of acetic acid. You can make these solutions by using serial dilution. The 0.010 M solution is made by obtaining 1.0 mL of 0.10 M acetic acid in a 10.0-mL graduated cylinder. It is important that you carefully adjust the amount with a dropper so that the meniscus reads exactly 1.0 mL. Then add distilled water to yield a total volume of 10.00 mL in the graduated cylinder. Pour the contents of the graduated cylinder into the small beaker. Measure the pH of this 0.010 M solution. Rinse the graduated cylinder several times with distilled water, then empty and dry it. Then put exactly 1.0 mL of the 0.010 M solution into the cylinder with a clean dropper to continue the dilution. Empty the contents of the beaker as your instructor directs. Clean and dry the beaker.

2-4. The 0.0010 *M* solution is made by combining 1.0 mL of the 0.010 *M* solution in the cylinder with 9.0 mL of distilled water, so that the total volume is 10.0 mL. Pour the contents of the graduated cylinder into the small beaker. Measure the pH of this 0.0010 *M* solution. Rinse the graduated cylinder several times with distilled water, then empty and dry it. Next, put 1.0 mL of the 0.0010 *M* solution into the cylinder to continue the dilution. Empty the contents of the beaker as your instructor directs. Clean and dry the beaker.

2-5. The 0.00010 *M* solution is made by a similar dilution. Record the pH. Empty the contents of the beaker as your instructor directs and clean all glassware.

Effects of Adding a Salt

3-1. Obtain a clean, dry, 50-mL beaker, a 10-mL graduated cylinder, and a dropper. Put exactly 5.0 mL of 0.10 *M* acetic acid into the beaker. Measure and record the pH.

3-2. Obtain two samples of sodium acetate—one about 0.050 g and another about 0.150 g. Record the mass exactly.

3-3. Add the sample of sodium acetate that is about 0.050 g to the beaker. Stir with a stirring rod. Measure and record the pH of the solution.

3-4. Add the sample of sodium acetate that is about 0.150 g to the beaker. Stir with a stirring rod. Measure and record the pH of the solution. Record the total mass of sodium acetate needed. Clean the beakers and dispose of the contents as your instructor directs. Have your instructor sign your notebook and Report Form.

24   **EXPERIMENT 24: ACIDS AND pH**

Prelab Exercises

1. What is the formula and molar mass of sodium acetate? Record this in your lab notebook.

2. If the pH is 5.83, what is the hydrogen ion concentration? (Express your answer using the correct number of significant figures.)

3. How many significant figures can be determined when obtaining ~9 mL of solution using a 10-mL graduated cylinder with markings at each mL and 0.1 mL? (If you don't know, give your best guess.)

4. Calculate the pH of a 0.0776 M H^+ solution. Express your answer in the correct number of significant figures.

Date _____ **Student's Signature** _____

24 EXPERIMENT 24: ACIDS AND pH

Report Form

DATA

Methods of Collecting Acid–Base Information

Give your observations for each below:

	HCl	HNO_3	NaOH	KHP	$H_2C_2O_4$
Red Litmus					
Blue Litmus					
PHN Indicator					
pH					
BTB Indicator					

	$HC_2H_3O_2$	KOH	Distilled Water	$Ba(OH)_2$	H_2SO_4
Red Litmus					
Blue Litmus					
PHN Indicator					
pH					
BTB Indicator					

pH and Concentration of an Acid

                                    pH

0.10 *M* acetic acid   _____

0.010 *M* acetic acid   _____

0.0010 *M* acetic acid _____

0.0001 *M* acetic acid _____

Effects of Adding a Salt

Formula of sodium acetate   _____

Formula mass of sodium acetate   _____

Mass of sodium acetate: sample 1 _____ g

Mass of sodium acetate: sample 2 _____ g

                        Total Sodium Acetate   *pH*

0.10 *M* acetic acid +   0.00 g   _____

0.10 *M* acetic acid + _____ g   _____

0.10 *M* acetic acid + _____ g   _____

Date _____**Instructor's Signature** _____

ANALYSIS

Methods of Collecting Acid–Base Information

What patterns do you see in the data? What substances react in a similar manner? List the substances that react the same to litmus, phenolphthalein, and bromothymol blue.

Based on the pH, which substances are acids and which are bases?

Explain how acids react with litmus, phenolphthalein, and bromothymol blue. Do likewise for bases.

Determine the hydrogen ion content of each of the acids based on the pH. Show calculations below.

Each acid was 0.10 *M*. Compare the hydrogen ion content of each. Are they constant or do they vary? Give an explanation for your findings.

Draw a particle view of the solution of acetic acid ($HC_2H_3O_2$ or CH_3COOH).

pH and Concentration of an Acid

From the pH, calculate the concentration of hydrogen ion. Next find the concentration of the acetate ion. (How would the concentration of hydrogen ion be related to the acetate ion? Hint: Look at the formula for acetic acid.) Next, determine the actual concentration of acetic acid that is in the solution. Show a set of sample calculations.

	pH	$[H^+]$	$[C_2H_3O_2^-]$	$[HC_2H_3O_2]$
0.10 *M* acetic acid	_____	_____	_____	_____
0.010 *M* acetic acid	_____	_____	_____	_____
0.0010 *M* acetic acid	_____	_____	_____	_____
0.0001 *M* acetic acid	_____	_____	_____	_____

Is the ionization of acetic acid a relatively complete reaction, or is it a partial one that represents an equilibrium?

What general pattern do you observe in the concentration of acetic acid versus pH?

Compare the relationship among the three values ($[H^+]$, $[C_2H_3O_2^-]$, and $[HC_2H_3O_2]$) for each initial concentration of acetic acid. Try different mathematical operations (x, /). You can try multiplying all three or a combination of multiplying and dividing (i.e., dividing one by the product of the other two, etc.).

Do these data give a constant or numbers having the same or nearly the same exponential values?

Write the equation for the ionization of acetic acid.

Effects of Adding a Salt

Calculate the number of moles of sodium acetate added. Then calculate the hydrogen ion concentration from the pH. What are the possible sources of acetate ion? Are sodium salts soluble? Calculate the concentration of acetate ion from all sources in the solution. Finally, calculate the concentration of acetic acid in the solution after mixing in the sodium acetate. All these solutions contained 1.00 M acetic acid initially. Use the space below the table to show your work for the first trial.

Sodium Acetate	pH	$[H^+]$	$[C_2H_3O_2^-]$	$[HC_2H_3O_2]$
0.00 mol	_____	_____	_____	_____
_____ mol	_____	_____	_____	_____
_____ mol	_____	_____	_____	_____

What is the relationship between acetic acid and sodium acetate? How are their formulas related?

What general pattern(s) do you see in the data?

Compare the relationship among the three concentrations. You can try multiplying all three or a combination of multiplying and dividing (i.e., dividing one by the product of the other two, etc.). Do the three sets of data give a constant or numbers that are similar? Is it similar to the relationship you found in the pH and Concentration section?

POSTLAB QUESTIONS

1. Compare the merits of the various methods for determining the acid–base nature of a substance.

2. Explain why the pH of every 0.01 M acid is not 2.0.

3. What is meant by a weak acid?

4. What is K_a? Does this have any relationship to this laboratory?

5. Explain what determines the pH of a weak acid.

6. What conclusion can be drawn about the effect of adding a conjugate salt on the pH of the solution?

7. Predict what should happen to the pH of an acetic acid solution if a salt that did not contain acetate ions were added.

Date _____ **Student's Signature** _____

Reactions of Acids and Bases

A Guided Inquiry Experiment

INTRODUCTION

Acids and bases are important classes of compounds. During a previous experiment, you investigated the pH of some acidic and some basic solutions. In that experiment, indicators were used to show acidity and basicity. In lecture, you have also studied the difference between strong and weak acids. In this experiment, you will look at some properties of acids, bases, and salts. You will study one reaction in greater detail.

OBJECTIVES

In this experiment, you will investigate the relationship between the reactants and products in an acid–base reaction. You will also study the effects of temperature and of the addition of another compound.

CONCEPTS

During this experiment, you will investigate the concepts of strong or weak acids, strengths of strong and weak bases, the formation of salts, and the titration of acidic and basic solutions.

TECHNIQUES

Titration, standardizing an alkaline solution, and using a buret are some of the techniques that you will use during this experiment.

ACTIVITIES

In this experiment, you will investigate the pH of solutions of salts. You will then investigate the reaction between a monoprotic, unknown, weak acid and a solution of NaOH. Based upon your observations, you will make conclusions about the concentration of the unknown weak acid and the nature of the acid–base reaction.

CAUTION

Use approved eye protection. Sodium hydroxide is a strong base; handle it carefully and avoid contact with your skin. If contact with any acid or base has occurred, wash with plenty of tap water.

PROCEDURES

pH of Salts

1-1. Obtain a spot test plate or a well plate and place it on a piece of white paper. Locate three spots that are evenly spaced over the length of the plate. Fill these three spots ¾ full of distilled water.

1-2. Place between 0.150 and 0.200 g of NaCl on a creased piece of smooth paper (~4 inches × 4 inches). Do likewise for $NaC_2H_3O_2$ and NH_4NO_3. Be sure to label the paper!

1-3. Add the crystals of NaCl to one spot. Stir with a toothpick. If a few crystals of NaCl do not remain undissolved, add more salt. Test the solution with pH paper. Record your results.

1-4. Repeat Procedure 1-3 for the other salts.

1-5. Add 2 to 4 drops of phenolphthalein indicator to each well. Record your results. Empty, clean, and rinse your spot plate.

Investigating the Reaction

2-1. Obtain about 2 mL of an unknown weak acid in a small test tube. Using a plastic pipet, put 10 drops of the weak acid into one spot of your plate.

2-2. Add 2 to 3 drops of phenolphthalein indicator. Record your results.

2-3. Obtain about 3 mL of a NaOH solution in a second small test tube. The NaOH solution should be about 0.1 *M*.

2-4. Rinse and dry the plastic pipet. Use the pipet to add one drop of the NaOH to the acid on the spot plate. Record your results.

2-5. Continue to add the NaOH until you observe a change. Record your results. Test the resulting solution with pH paper. Record the results.

2-6. Dispose of the solution as your instructor indicates. Clean and rinse your pipettes and plate.

Standardizing the NaOH

3-1. In your Prelab Exercises, you calculated the mass of potassium hydrogen phthalate needed to neutralize 10 mL of an ~0.1 *M* NaOH solution. Have your instructor check your calculations.

3-2. Tare a clean, creased piece of smooth paper (~4 inches × 4 inches) on the analytical balance. Carefully add to the paper approximately the mass of potassium hydrogen phthalate (KHP) needed to neutralize 10 mL of a 0.1 *M* NaOH solution. The mass actually obtained should not be exactly the amount you needed. ***Remember that the balance must be clean when you leave it.*** Record the mass of your sample. Transfer your sample of KHP to a clean but not necessarily dry 250-mL Erlenmeyer flask.

3-3. Add approximately 100 mL distilled water to the flask. Carefully swirl and gently warm the flask until all the phthalate has dissolved. Then remove the flask from the flame and allow it to partially cool.

3-4. Add three or four drops of phenolphthalein indicator to the KHP solution in the flask and set the flask on a sheet of white paper under the buret.

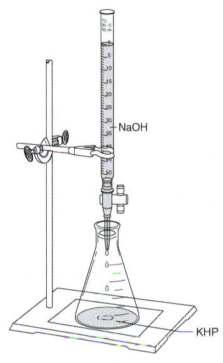

3-5. Clean and fill a buret with the NaOH solution that is about 0.1 *M*. In your notes, record the volume in the buret to the nearest 0.01 mL. Record the volume. While gently swirling the warm flask, slowly add the ~0.1 *M* NaOH solution to the KHP solution until a pale pink endpoint is observed and which persists for about a minute. Record the final buret reading. Empty the flask and rinse it with distilled water. Repeat Procedures 3-1 through 3-5 for Trial 2. Keep the buret for the next section.

Quantitative Measurements

4-1. Obtain a pH meter. Calibrate the pH meter using the buffer solutions provided. Your instructor will provide instructions for the type of pH meter to be used.

4-2. Obtain 20 mL of unknown weak acid in a 250-mL beaker from the buret your instructor has set up in the room. Add 3 to 4 drops of phenolphthalein and 30 mL of distilled water.

4-3. Place the beaker on a stirrer with the switch off. Then, carefully insert a magnetic stir bar into the solution, turn the speed to slow, and turn on the switch. Measure the pH of the solution. Record this as the initial pH.

4-4. Add the standardized NaOH dropwise from the buret at your station. Record the pH and volume of NaOH when there is a **change of 0.15 to 0.25 pH units.** Record the volume when the endpoint is reached.

4-5. Continue until you have recorded 4 to 5 readings above the pH 12.

Repeat Procedures 4-1 through 4-5 with a second trial.

4-6. Dispose of the contents as your instructor directs. Clean and return your equipment. Have your instructor sign your notebook and Report Form.

25   EXPERIMENT 25: REACTIONS OF ACIDS AND BASES

Prelab Exercises

1. Calculate the mass of potassium hydrogen phthalate needed to neutralize 10 mL of an ~0.1 M NaOH solution. Record this value in your notebook. (Consult your textbook regarding the reaction of KHP with a base.)

2. Give the general reaction that occurs between an acid and a base.

3. What is a salt?

4. Explain why the NaOH must be standardized.

5. Describe the difference between equivalence point and endpoint.

6. Answer any question assigned by your instructor.

Date _____ **Student's Signature** _____

Name (Print) | Date (of Lab Meeting) | Instructor

Course/Section | Partner's Name (If Applicable)

25 EXPERIMENT 25: REACTIONS OF ACIDS AND BASES

Report Form

DATA

pH of Salts

Give your observations for pH and phenolphthalein below:

$NaCl$

$NaC_2H_3O_2$

NH_4Cl

Investigating the Reaction

Give your observations when phenolphthalein was added to the weak acid:

Give your observations when NaOH was added dropwise to the weak acid:

Describe your observations when additional NaOH produced a change in the color of the indicator and a change in the pH.

Standardizing the NaOH

Mass of KHP calculated to neutralize 10 mL of ~0.1 M NaOH: _____ g

Phthalate titration data

	Trial 1	Trail 2		Trial 1	Trial 2
Phthalate	_____ g	_____ g	Final volume NaOH (at the endpoint)	_____ mL	_____ mL
			Initial volume NaOH	_____ mL	_____ mL
			Vol. NaOH consumed	_____ mL	_____ mL

Quantitative Measurements

Record your pH and volume data in your lab notebook. (Be sure to include the volume at the point at which the indicator changed color.)

Date _____ Instructor's Signature _____

ANALYSIS

pH of Salts

Are the following salts acidic, basic, or neutral according to your results?

NaCl _____

$NaC_2H_3O_2$ _____

NH_4NO_3 _____

Describe the strength (strong or weak) of the parent acid and the parent base that produced the following salts:

	Parent Acid	Parent Base
NaCl	_____	_____
$NaC_2H_3O_2$	_____	_____
NH_4NO_3	_____	_____

Can you make a generalization about the acidity of a salt and its parents (or conjugates)?

Investigating the Reaction

Draw a particle view of the reactants and products in the reaction between NaOH and the weak monoprotic acid (HA).

Standardizing the NaOH

	Trial 1	Trial 2

Standardized concentration of NaOH solution: _____ *M* and _____ *M*

Average standardized concentration of NaOH solution: _____ *M*

Quantitative Measurements

Attach a graph of volume of pH vs. NaOH. On the graph, construct lines to determine the equivalence point. If you are not sure about a method to accomplish this, check the Common Procedures and Concepts Section at the end of this manual or with your instructor for details.

You began with an acid in the beaker and slowly added a base. What substance(s) were in the beaker after the initial point, but before the equivalence point?

What substance(s) were in the beaker at the equivalence point?

What substance(s) were in the beaker after the equivalence point?

Using the more accurate equivalence point rather than the endpoint, how many mL of NaOH were required to neutralize all of your weak acid?

A good indicator choice gives a color change (endpoint) at or very near the equivalence point. Did we make a good indicator choice? Explain why or why not.

What is the molarity of the unknown acid? (Assume that it is a weak, monoprotic acid.) Explain how you determined this.

A buffer can be made from a weak acid and a salt of the weak acid. Did you form a buffer at any point during the quantitative measurement part of the experiment? If so, describe when and mark this section of your graph.

POSTLAB QUESTIONS

1. You know that weak acids exist in aqueous solutions in equilibrium with their dissociated ions. Write the equilibrium expression as a function of the K_a of a weak acid.

$K_a =$

2. If ½ of the weak acid were neutralized into salt, how would the concentration of weak acid compare to the concentration of the salt? (This is called the ½ neutralization point.)

3. If a sample is at the ½ neutralization point, what species are equivalent and can be canceled from the K_a equation?

Write the new equilibrium expression after canceling equivalent species.

$K_a =$

4. How many mL of NaOH were required to neutralize ½ of your weak acid? _____ (You found the number to neutralize all of your weak acid in the Analysis section.)

What was the pH at the ½ neutralization point? _____ Mark this on your graph.

What is the K_a for your weak acid?

What is the pK_a for your weak acid?

5. From the appendix in your textbook, what do you think the identity of your weak acid was?

6. What is the relationship between pH and pK_a at the ½ neutralization point? Is that relationship true at points other than the ½ neutralization point?

Date _____   **Student's Signature** _____

Acids and Bases

An Open Inquiry Experiment

INTRODUCTION

Acids and bases are important classes of compounds. The interaction of acids and bases can be investigated using indicators, the measurement of heat given off during a reaction, or with an instrument such as the pH meter, etc. In some experiments, you are directed as to which data to collect and which chemicals to use; however, in this experiment, you will design the exact experimental procedures. This type of lab is often referred to as an open inquiry lab. "Open" means that you design the procedures. In addition, you also have a number of options from which you can choose the chemical system you want to investigate.

OBJECTIVES

In this experiment, you will design an experiment that will allow you to determine specific values, depending on the option you choose to investigate.

CONCEPTS

This experiment uses the following concepts: solution concentration, standardization, titration, pH, neutralization, and pH at the half-neutralization point.

TECHNIQUES

Using a buret and volumetric glassware are two of the techniques encountered in this experiment.

ACTIVITIES

You will design an experiment that will provide information about the interactions of acids and bases, including the acid neutralizing capacity of an antacid tablet, the heat of neutralization, and the titration of a weak acid/strong base or weak base/strong acid.

CAUTION

You will be working with dilute solutions but you should treat all solutions as potentially dangerous. Dispose of materials as directed by your instructor. Wear approved eye protection.

PROCEDURES

Experimental Design

1-1. Design an experiment from the options in 1-3.

The design must include:

Problem Statement: This includes a few sentences describing the specific question(s) you are trying to answer with your experiment.

Proposed Procedures: This section contains the materials and equipment that you will use, the type of data you will collect (the variables you will measure), and the number of trials you are proposing. Remember to discuss safety considerations. Your planned experimentation should take up ⅔ of the lab period.

1-2. Remember that multiple runs will establish the reliability of your data. Try to keep the amounts small and the concentrations dilute. Also, it is important to standardize acids and bases. Unless indicated otherwise, all stock solutions will be provided as 0.10 *M* solutions.

1-3. Choose from the following four options:

Option 1

Investigate the acid neutralizing capacity of an antacid using the pH meter. Do this by adding acid to a sample of antacid until pH = 2 is obtained and then titrating the excess acid with standardized NaOH. Remember that the pH of the stomach is 3.5. Choose one of the following acids:

HCl

HNO_3

(It is best to use small samples of the antacid tablet; for example, 0.2 g or less.)

Option 2

Investigate the neutralization point of a titration using indicators. Then prepare a solution at the half-neutralization point of the acid or base using universal indicator and equate the resulting color to a pH. Determine the K_a or K_b of the acid or base from your experimental data. Select from the following (a suggested indictor is in parenthesis):

NH_3 titrated with 0.1 *M* HCl (methyl orange)

$C_3H_6O_2$ titrated with 0.1 *M* KOH (phenolphthalein)

NH_3 titrated with 0.1 *M* HNO_3 (methyl orange)

Option 3

Investigate the heat released during neutralization. Determine the amount of heat released per mole of acid. Also, determine the relationships among the amount of heat released, the amount of acid (concentration of the acid), and the species of the acid. Use ~2 *M* NaOH as the base. Both 1.0 *M* and 0.1 *M* acids will be available. You may choose from the following:

$HC_2H_3O_2$, H_2SO_4, HCl, HNO_3

Option 4

Investigate an aspect of acid–base interaction with a chemical system. Be sure that your instructor approves.

1-4. Your instructor will supply feedback on your experimental design BEFORE you do the experiment.

Experimentation

2-1. Make any changes that your instructor suggests, and then proceed to collect data. You will collect data with a partner and should discuss your results with your partner. These discussions will help you both learn.

2-2. Record the data in your notebook and get your instructor's signature when all data is collected.

Report

3-1. After completing the experiment, you will write a lab report. Although you collect data and share ideas with a partner, you will be expected to write the final lab report independently. Your grade will depend on the thoroughness of your investigation, the presentation of your data, the careful analysis of the data, and the logic used to give reasonable results and explanations.

3-2. The lab report MUST include the following four sections:

Problem Statement: This includes a few sentences describing what specific questions you are trying to answer with your experiment.

Procedures: This section contains the materials and equipment that you actually used (these may differ from those you proposed), the type of data collected (the variables measured), and the number of trials done. Remember to discuss safety considerations. Your experimentation should take up ⅔ of the lab period.

Data/Analysis: Include the data you collected. Data should be in tables when possible, with easy-to-read labels. Analysis of the data should also be included (analysis = what your data tells you). Graphs (with labels, units, and titles), mathematical relationships, chemical equations, and algebraic equations should be given, and the connection to the data should be shown.

Conclusion: This is the generalization or explanation you have deduced from your experiment. This is also the place to make explanations for any data results that are counter to logical chemical ideas and to describe how you would change the experiment if you repeated it.

Name (Print) _____   Date (of Lab Meeting) _____   Instructor _____

Course/Section _____

26   EXPERIMENT 26: ACIDS AND BASES

Prelab Exercises

Problem Statement

Proposed Procedures (insert more sheets if needed)

Date _____   **Student's Signature** _____

Instructor's Approval and Comments:

Date _____ **Instructor's Signature** _____

26   EXPERIMENT 26: ACIDS AND BASES

Report Form

DATA

Data is collected in your notebook.

Date _____ **Instructor's Signature** _____

LAB REPORT

Attach this sheet to your lab report that includes the PROBLEM STATEMENT (actual), PROCEDURES (actual), DATA/ANALYSIS, and CONCLUSION.

POSTLAB QUESTIONS

1. Draw a particle view(s) of the reactants and products in the reaction(s) in your experiment.

2. What are the most important factors to keep in mind when designing an experiment? Discuss at least three.

3. Answer any questions assigned by your instructor.

Date _____ **Student's Signature** _____

Electrochemistry

A Guided Inquiry Experiment

INTRODUCTION

Electrochemistry is a branch of chemistry that deals with the production of electricity by chemical reactions and the changes produced by electrical current. You have probably seen potato clocks, lemon clocks, or similar devices that produce electricity via a chemical reaction. Batteries are a good example of this topic. Passing an electrical current through a solution can cause metals to plate out or gases to be formed. These are examples of electrolysis caused by an electrical current. In this experiment, you will investigate reactions that produce electricity.

OBJECTIVES

In this experiment, you will investigate the relationship between the reactions of metals and their ions (half-cells), and voltages produced when two half-cells are combined to form electrochemical cells. You will also explore the effects that changes in concentration can have on the voltage produced.

CONCEPTS

This experiment uses the concepts of solution concentration and reactions. Redox reactions are used throughout this experiment.

TECHNIQUES

Diluting a solution, reading a multimeter, and graphing experimental data are some of the techniques that you will use during this experiment.

ACTIVITIES

In this experiment, you will collect data concerning the reactions of metals and their ions, any voltages produced by combinations of half-cells constructed of metals and their ions, and voltages produced by half-cell combinations of differing concentration.

CAUTION

Use approved eye protection. Be aware of chemicals and hot equipment.

PROCEDURES

Do Metals & Metal Ions React?

1-1. Obtain a well plate, four rectangular pieces (strips) of metal (Cu, Pb, Zn, Mg), and steel wool or sandpaper.

1-2. Clean the four metal strips by using sandpaper or steel wool. The metals MUST also be thoroughly recleaned BEFORE and AFTER each use.

1-3. Place the well plate on a sheet of paper on which you can make notes about the contents of each well. Fill four wells about ¾ full with 0.10 M Cu^{2+}. Do likewise with the other 0.10 M metal ion solutions (Pb^{2+}, Zn^{2+}, and Mg^{2+}). Sixteen wells will be about ¾ full when this step is completed.

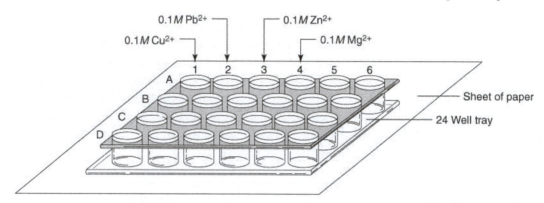

1-4. Put one of the four metal strips into each well containing the Cu^{2+} solution. Allow these to set for a few minutes. A reaction may appear as a black residue on the metal strip. You can wipe the metal strip on a paper towel to see if there is any residue. Record your observations; use NR for no reaction. Remove the metal strips, clean them as before, and repeat the procedure with the Mg^{2+} solution, then the Zn^{2+} solution, and finally the Pb^{2+} solution.

1-5. When you have finished, empty the well plate as directed by your instructor. Rinse the well plate and invert it. Reclean the four metal strips with sandpaper or steel wool.

Measuring Voltage

2-1. On the well plate, fill one well about ¾ full of the Zn^{2+} solution. Place the Zn strip into the Zn^{2+} solution. You have just created a Zn half-cell (a metal in a solution of its ions). This half-cell is indicated by the symbols: $Zn|Zn^{2+}(0.1\ M)\|$. Do likewise with each of the three metal ion solutions and metal strips to create Pb, Cu, and Mg half-cells.

2-2. Connect the Zn and Cu half-cells with a "salt bridge." You can create a salt bridge by rolling a small piece of paper towel (about 3 cm × 9 cm) into a rope and soaking it in $NaNO_3$ or in NH_4NO_3. Another method for creating a salt bridge is to use a strip of filter paper on which you have squirted a continuous line of $NaNO_3$ or NH_4NO_3. Place one end of your salt bridge into each half-cell.

2-3. Your instructor will give you instructions on how to use your particular brand of multimeter. For most brands, the black lead goes in the common; the red lead in the + slot. Set on +DC 2.50 V. Read the scale that goes from 0 to 250, but move the decimal over two places to the left.

2-4. Touch the leads of the multimeter to the Zn and Cu metals of the half-cells connected with the salt bridge. Any time you find a negative voltage, reverse the leads. Clamp the leads onto the metal strips. This set-up is called a voltaic cell. Read the voltage of this voltaic cell. THE LEADS MUST BE CLAMPED TO THE METALS TO GET GOOD READINGS.

2-5. Make a new combination of half-cells. If you are using filter paper, use a new salt bridge with each combination; if you are using string, rinse the ends of the string in a watch glass of water and then soak the string in a solution of $NaNO_3$ or NH_4NO_3 before using it with a new combination of half-cells. Record the voltage. Continue until you have measured the voltage of all six combinations of the four hal-cells.

2-6. Now repeat the whole procedure to obtain a second voltage reading for each combination. Calculate an average reading for the voltage for each combination of half-cells.

2-7. When you have finished, empty the well plate as your instructor directs. Rinse it and invert it. Clean the end of the four metal strips with sandpaper or steel wool.

Concentration Effects

3-1. Fill one well on the well plate about ¾ full of 0.10 M Zn^{2+} solution. Place Zn metal strip in the Zn^{2+} solution. Fill a well on the well plate about ¾ full of the 0.10 M Cu^{2+} solution, but do not put in the Cu metal strip.

3-2. You will need to prepare solutions for three more Cu half-cells. You will need half-cells made with 0.010 M Cu^{2+}, 0.0010 M Cu^{2+}, and 0.00010 M Cu^{2+}. You can make these solutions by using serial dilution. The 0.010 M solution is made by obtaining 1.0 mL of 0.10 M Cu^{2+} in a 10.0-mL graduated cylinder. Carefully reading the bottom of the meniscus, add distilled water until there are exactly 10.00 mL in the graduated cylinder. Mix the solution with a stirring rod. Fill a well on the well plate about ¾ full of the 0.010 M Cu^{2+} solution. Save 1.0 mL of this solution in the graduated cylinder for preparation of the next solution.

3-3. The 0.0010 M solution is made by diluting 1.0 mL of the 0.010 M Cu^{2+} solution to a total volume of 10.00 mL. The 0.00010 M Cu^{2+} solution is made likewise. Fill a well about ¾ full of each of the Cu^{2+} solutions.

3-4. Clean your copper strip and place it in the 0.00010 M Cu^{2+} solution. Connect this half-cell to the Zn half-cell with a new salt bridge. Connect the voltmeter and read the voltage.

3-5. Remove the Cu strip and the salt bridge. Rinse or replace both. Place the Cu strip into the 0.001 M Cu^{2+} solution and connect this half-cell to the Zn half-cell with a salt bridge. Connect the voltmeter and read the voltage. Repeat this procedure with the 0.010 M Cu^{2+} solution and the 0.10 $M\,Cu^{2+}$ solution.

3-6. Now repeat the steps above to obtain a second voltage reading for each. Calculate an average reading for the voltage for each combination of half-cells. Remove the Zn metal from the Zn solution, but leave the Cu metal in the 0.10 M Cu^{2+} solution. Return the Zn metal strip to its original container.

3-7. Fill one well ¾ full with 0.25 M Cu^{2+} solution. Obtain a second strip of Cu metal. Clean both copper metal strips. Add one clean strip of Cu metal to the 0.0001 $M\,Cu^{2+}$ solution. Add the second clean strip of Cu metal to a 0.25 M Cu^{2+} solution. Connect the two solutions with a salt bridge. Measure the voltage of a cell made from 0.00010 $M\,Cu^{2+}$ solution and 0.25 M Cu^{2+} solution. Disconnect the voltmeter and repeat this a second time. Average the two readings. Remove the Cu metal from the 0.00010 M Cu^{2+} solution. Pat the metal dry with a paper towel.

3-8. Repeat Procedure 3-7 using the 0.001 M Cu^{2+} solution in place of the 0.0001 $M\,Cu^{2+}$ solution.

3-9. Repeat Procedure 3-7 using the 0.01 M Cu^{2+} solution in place of the 0.0001 $M\,Cu^{2+}$ solution.

3-10. Disassemble your cells. Rinse and dry the metals. Dispose of the solutions as directed by your instructor. Have your instructor sign your notebook and Report Form.

27   **EXPERIMENT 27: ELECTROCHEMISTRY**

Prelab Exercises

1. Explain how 10.0 mL of a 0.100 M solution of copper ions is prepared from solid $Cu(NO_3)_2$.

2. What is meant by the symbols: $Cu|Cu^{2+}(0.1\ M)$?

3. What is a redox reaction? Give an example of a redox reaction.

4. Answer any questions assigned by your instructor.

Date _____ **Student's Signature** _____

27 EXPERIMENT 27: ELECTROCHEMISTRY

Report Form

DATA

Do Metals & Metal Ions React?

Give your observations for the combinations of metals and metal ions below:

	Cu	Mg	Zn	Pb
Cu^{2+}				
Mg^{2+}				
Zn^{2+}				
Pb^{2+}				

Measuring Voltage

	Voltage Reading #1	Voltage Reading #2	Average
$Zn|Zn^{2+}||Cu^{2+}|Cu$			
$Zn|Zn^{2+}||Pb^{2+}|Pb$			
$Mg|Mg^{2+}||Pb^{2+}|Pb$			

	Voltage Reading #1	Voltage Reading #2	Average
$Pb|Pb^{2+}||Cu^{2+}|Cu$			
$Mg|Mg^{2+}||Cu^{2+}|Cu$			
$Mg|Mg^{2+}||Zn^{2+}|Zn$			

Concentration Effects

	Voltage Reading #1	Voltage Reading #2	Average
$Zn|Zn^{2+}||\ 0.00010\ M\ Cu^{2+}|Cu$			
$Zn|Zn^{2+}||\ 0.0010\ M\ Cu^{2+}|Cu$			
$Zn|Zn^{2+}||\ 0.010\ M\ Cu^{2+}|Cu$			
$Zn|Zn^{2+}||\ 0.10\ M\ Cu^{2+}|Cu$			

	Voltage Reading #1	Voltage Reading #2	Average
$Cu|0.00010\ M\ Cu^{2+}||\ 0.25\ M\ Cu^{2+}|Cu$			
$Cu|0.0010\ M\ Cu^{2+}||\ 0.25\ M\ Cu^{2+}|Cu$			
$Cu|0.010\ M\ Cu^{2+}||\ 0.25\ M\ Cu^{2+}|Cu$			

Date _____   **Instructor's Signature** _____

ANALYSIS

Do Metals & Metal Ions React?

Write the balanced chemical equations for each reaction observed on an attached sheet. For two of the equations, explain why you chose the particular products you did.

What trends do you see in the relative reactivity of the metals in **DO METALS & METAL IONS REACT?**

What trends do you see in the relative reactivity of the metal ions in **DO METALS & METAL IONS REACT?**

How does the tendency of the metals to lose electrons compare with the tendency of the metal ions to gain electrons?

Draw a particle representation of the reaction between the Mg and the Zn^{2+}.

Measuring Voltage

If electricity is a flow of electrons, choose one combination of cells to explain how the flow of electrons travels.

What reactions are happening at each of the half-cells?

How do the reactions of the two half-cells combine to cause a flow of electrons?

What is the net reaction of the two half-cells?

What are the relationships among the voltages of the $Zn|Zn^{2+}||Cu^{2+}|Cu$ cell, the $Zn|Zn^{2+}||Pb^{2+}|Pb$ cell, and the $Pb|Pb^{2+}||Cu^{2+}|Cu$ cell? Explore the relationship of the similarly related cells such as the $Mg|Mg^{2+}||Cu^{2+}|Cu$ cell, the $Mg|Mg^{2+}||Pb^{2+}|Pb$ cell, the $Pb|Pb^{2+}||Cu^{2+}|Cu$ cell, etc.

Choose one of the voltaic cells to draw. Show how the electrons and ions are moving through the cell. Label all of the parts.

Concentration Effects

What conclusion can be drawn from the voltages of the Zn half-cell with the Cu half-cells of differing concentrations? What general pattern exists when the concentration is changed?

Graph the data to find the numerical pattern. Is the relationship linear? How do you know? What other mathematical patterns are possible? Attach graph(s) to this lab.

Can two half-cells containing the same ion but at different concentrations create a potential (voltage)?

Which combination of concentrations of the same ion gave the greatest voltage?

What trend in the concentration difference of Cu ions versus voltage did your data reveal?

POSTLAB QUESTIONS

1. If you were manufacturing a battery from your half-cells, describe a design that would give the largest voltage possible from these materials.

2. What is the activity series?

Date _____ **Student's Signature** _____

Small-Scale Redox Titration

A Skill Building Experiment

INTRODUCTION

In many laboratories, technicians are called upon to determine the concentration of an ion or compound in solution. Solutions of acids, bases, oxidizing agents, and reducing agents for which the concentration of the solute must be known precisely and accurately are needed as standards in many of these determinations. As an example, laboratories associated with water quality may wish to determine: (1) the degree of hardness of a water sample by reacting it with a solution of disodium ethylenediamine tetraacetate (EDTA), (2) the sample's chemical oxygen demand by titration with a potassium dichromate solution, (3) its alkalinity by titration with a solution of an acid, or (4) its acidity by titration with a solution of a base. Many analytical procedures are now automated, but the time-honored titration is still sometimes the method of choice.

OBJECTIVES

In this experiment, a solution of $KMnO_4$ is prepared, standardized, and used to determine the concentration of an unknown iron solution. From given information about how the iron solution had been prepared from a known mass of an iron ore plus the solution's concentration, the percentage of the iron in the ore is to be determined.

CONCEPTS

An aqueous solution of potassium permanganate will react with an acidified aqueous solution of sodium oxalate according to the following equation. This redox reaction is catalyzed by Mn^{2+} ion, a product of the reaction:

$$2KMnO_4 + 5Na_2C_2O_4 + 8H_2SO_4 \xrightarrow{Mn^{2+}} 2MnSO_4 + 10CO_2 + K_2SO_4$$
$$+ 5Na_2SO_4 + 8H_2O$$

During titrations, such as those being run in this experiment, a solution of known concentration (the titrant) is added to a sample until the equivalence point is reached. At the equivalence point, the titrant has reacted all of the substance to be determined in the sample (the analyte).

The concentration of analyte in the solution can then be calculated from the volume of the titrant consumed, the concentration of the titrant solution, the mass of the analyte used, and the mole ratio between titrant and analyte.

$$M_{titrant} \times V_{titrant} = Moles_{titrant}$$

$$Moles_{titrant} (\text{``b''} \, moles_{analyte} / \text{``a''} moles_{titrant}) = Moles_{analyte} \text{ in solution}$$

(``b'' and ``a'' are the coefficients of the analyte and titrant in the balanced equation.)

$$Moles_{analyte} = (mass_{analyte}/formula \ weight_{analyte})$$

To obtain reliable results, the concentration and volume of the titrant and the volume of analyte must be accurately known. Solutions of accurately known concentrations generally cannot be made by weighing a required amount of the titrant and dissolving it in water. Many potential titrants cannot be obtained in sufficiently high purity or (like in the case of sodium hydroxide) they react with the carbon dioxide and/or water in the air. Still others, such as potassium permanganate, react or deteriorate on standing. Fortunately, a few substances can be obtained in high purity, are stable in the solid state and in solution, and do not deteriorate on contact with air. These compounds are called primary standards and are used to prepare primary standard solutions with concentrations that are precisely and accurately known. The standardized solutions are then employed to analyze a variety of samples. Potassium hydrogen phthalate (KHP) is an example of a primary standard for base solutions, and sodium oxalate is an example of a primary standard for potassium permanganate solutions.

In the above equation, the reaction is catalyzed by Mn^{2+}, a product of the reaction. Since Mn^{2+} ion is not present in the solution at the beginning of the reaction, $MnSO_4$ must be added before the reaction begins. The endpoint of this reaction is reached when one drop of $KMnO_4$ solution is not decolorized anymore and imparts a light-pink color to the analyte solution. When a $KMnO_4$ solution is to be standardized, the required amount of pure sodium oxalate is weighed to ±0.1 mg, dissolved in water, acidified with sulfuric acid containing a catalytic amount of Mn^{2+}, and titrated with the $KMnO_4$ solution to the endpoint. The number of moles of oxalate can be calculated from the mass of sodium oxalate used and its formula weight. From the titration data, the molarity of the $KMnO_4$ solution can be calculated. This then becomes a standardized $KMnO_4$ solution.

Once the $KMnO_4$ solution has been standardized and, thus, its concentration precisely and accurately known, the concentration of an iron(II) solution can be determined via titration. From the concentration of iron(II) ion in solution and a history of how the solution was prepared, one can calculate the amount of ion in the ore sample from which the iron solution may have been prepared.

TECHNIQUES

This experiment familiarizes you with the standardization of a potassium permanganate solution and the use of the standardized solution for the determination of the percentage of iron in an unknown. There are many similarities between the redox stoichiometry and acid–base stoichiometry.

ACTIVITIES

You are to prepare and standardize a potassium permanganate solution and use the solution to titrate an iron(II) solution. From your titration data and a history of how the iron(II) solution may have been prepared, you are to calculate the percentage of iron in the ore used to prepare the iron(II) solution. To gain confidence in the results, the standardization and the titration of the unknown are carried out several times. The values are averaged, and the standard deviation and the relative standard deviation are calculated. Under the conditions at which this experiment is performed, relative standard deviations of 1% or less than 1% are characteristic of precise results. Precision is not necessarily associated with accuracy.

CAUTION	

You must exercise proper care in heating and in transferring hot equipment and reagents. Dispose of materials as directed by your instructor. Wear approved eye protection at *all* times. Avoid contact of the $KMnO_4$ with organic material or concentrated H_2SO_4. $KMnO_4$ may produce an explosive mixture with organic material or concentrated H_2SO_4. If any of the reagents used in this experiment come in contact with your skin, wash the affected areas immediately and thoroughly with water. Notify your instructor. The reagents do not present a hazard unless they are misused.

PROCEDURES

Calibration of a Beral Pipet

1-1. Obtain distilled water in a 100-mL beaker. Find the number of drops per 1 mL for a plastic long-stemmed dropper (a Beral pipet). This can be done by counting the drops to fill a 10-mL graduated cylinder from the 6-mL mark to the 9-mL mark and dividing by 3. Repeat the determination twice more and average the results. Be sure to read the bottom of the meniscus.

Preparation of the KMnO₄ Solution

2-1. If $KMnO_4$ solution is provided as a more concentrated solution than and as a solution whose concentration has probably changed since it was first prepared, prepare 40 mL of a solution that is approximately 0.02 *M* $KMnO_4$ by diluting a volume of the more concentrated solution. Stir thoroughly. Skip to Procedure 3-1.

2-2. If the $KMnO_4$ is provided as a solid, place a clean, dry watch glass on the top-loader balance and tare the balance. Carefully add the crystals of $KMnO_4$ to the watch glass until you have added a mass of $KMnO_4$ approximately equal to that calculated for the preparation of 40 mL of 0.020 *M* $KMnO_4$ solution. Record the mass of the $KMnO_4$ to the nearest 0.01 g.

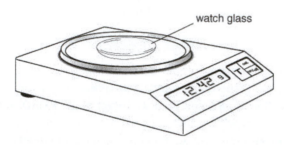

watch glass

2-3. Place 20 mL distilled water in a small beaker (100 mL or larger). Quantitatively transfer the $KMnO_4$ crystals into a 150-mL beaker. Use some of the water to rinse the last crystals off the watch glass and into the beaker. Pour the rest of the water directly into the beaker. Swirl the deep-purple solution for 5 minutes. Let the solution stand for a few minutes. Then decant the solution slowly into another large flask or beaker. Do not agitate the solution during this process. When the last milliliters of the solution are transferred, check for undissolved $KMnO_4$ crystals in the beaker from which you are pouring. If undissolved crystals are seen, return about 10 mL of the solution to the original flask and swirl until dissolution is complete. Combine the solutions. Swirl and stir.

The homogeneity of the KMnO₄ solution, i.e., the same concentration throughout the solution, is very important for the success of this experiment.

Standardization of the KMnO₄ Solution

3-1. Using the analytical balance and weighing paper, weigh an amount of sodium oxalate within ±5 mg of the quantity calculated to react with 2.0 mL of a 0.02 *M* KMnO₄ solution.

weighing paper

CAUTION

Sodium oxalate is toxic if ingested.

Record the mass of sodium oxalate to the nearest 0.1 mg (0.0001 g). Transfer the weighed sodium oxalate to a clean but not necessarily dry 125-mL Erlenmeyer flask. Use 30 mL of distilled water to dissolve the sodium oxalate. It may not all dissolve until it is heated in the next step.

3-2. Add 20 drops of 9 M H_2SO_4 to the sodium oxalate water mixture. Protect your thumb and fingers from the hot flask. Over a flame, warm the flask while swirling it until its temperature is approximately 70°C. When all the sodium oxalate has dissolved, set the hot flask on a sheet of white paper.

3-3. To your sodium oxalate solution add 2 drops of the 0.1 M $MnSO_4$ solution provided by your instructor. Drain the water from your calibrated pipet. Rinse it with a few drops of the $KMnO_4$ solution. Expel these drops into a waste beaker, then fill the pipet about ½ full of the $KMnO_4$ solution. Begin to titrate by adding one drop of $KMnO_4$ solution to the hot oxalate solution. Swirl until the solution has become colorless. Then add the next drop.

Keep a record of the drops added. As the Mn^{2+} concentration builds up, the reaction speeds up and you can add the titrant faster. If the solution becomes brownish, you added $KMnO_4$ too quickly or the solution cooled too much (or you didn't add the sulfuric acid). If necessary, reheat the solution. Titrate the sodium oxalate solution to the very pale-pink endpoint. Record the total number of drops of $KMnO_4$ solution used.

3-4. Repeat the standardization twice more either at this time or after the completion of your work with the unknown iron solution.

Titration of an Fe(II) Solution Like One Prepared From Hematite (an Ore Containing Fe₂O₃)

4-1. Using a buret provided by your instructor, read the volume of the Fe(II) solution provided in the buret to the nearest 0.01 mL. Drain approximately 1.2 mL of the Fe(II) solution provided into a clean 125-mL Erlenmeyer flask. Then read and record the liquid level in the Fe(II) buret.

4-2. Add 20.0 mL distilled water to your solution of Fe(II) in the 125-mL Erlenmeyer flask. Swirl the flask to mix the contents. Add 1 mL of 9 M H_2SO_4, two drops of the $MnSO_4$ solution provided by your instructor, and one drop of 3 M H_3PO_4. Swirl the flask again to mix its contents.

4-3. Place the Erlenmeyer flask on a piece of white paper. Refill your calibrated pipet if needed. Using your calibrated pipet, slowly add the $KMnO_4$ solution to the swirled Fe(II) solution. Keep a record of the drops of $KMnO_4$ solution added. The additions can be fairly rapid as long as the $KMnO_4$ solution is quickly decolorized. When the color fades only slowly or persists, stop the addition of $KMnO_4$ and allow the $KMnO_4$ to react. When the mixture has become colorless, continue the titration dropwise. The closer your mixture is to the endpoint, the slower you should add the $KMnO_4$. The endpoint is reached when one drop of the $KMnO_4$ solution gives your mixture a faint pink color that persists in the swirled mixture for 30 seconds. Record the total number of drops of $KMnO_4$ solution added.

4-4. Clean your 125-mL Erlenmeyer flask. Repeat the titration twice more by repeating Procedures 4-1 through 4-4. Refill your calibrated pipet as needed.

4-5. Place your unused reagents in designated collection containers and wash all equipment twice to ensure that all reactants are removed. Have your instructor sign your notebook and Report Form.

Name (Print) | Date (of Lab Meeting) | Instructor

Course/Section

28 EXPERIMENT 28: SMALL-SCALE REDOX TITRATION

Prelab Exercises

1. Define:

meniscus

solution

standardization

primary standard

2. Balance the following half-reactions:

$MnO_4^- \quad + \quad H^+ \quad \rightarrow \quad Mn^{2+} \quad + \quad H_2O$

$Na_2C_2O_4 \quad \rightarrow \quad CO_2 \quad + \quad Na^+$

3. Balance the following equations: (Record your answers in your notebook.)

a. $MnO_4^- \quad + \quad H_2C_2O_4 \quad + \quad H^+ \quad \rightarrow \quad Mn^{2+} \quad + \quad CO_2 \quad + \quad H_2O$

b. MnO_4^- + Fe^{2+} + H^+ → Mn^{2+} + Fe^{3+} + H_2O

4. Calculate the mass of $KMnO_4$ required for the preparation of 40 mL of 0.020 M $KMnO_4$ solution. (Record your answer in your notebook.)

5. Calculate the amount of sodium oxalate required to consume 2.0 mL of 0.020 M $KMnO_4$. (Record your answer in your notebook.)

6. Calculate the grams of iron ion present in 1.25 mL of a solution that required 1.93 mL of 0.02131 M $KMnO_4$ in a titration going to a faint pink endpoint. (Record your answer in your notebook.)

Date _____   **Student's Signature** _____

28   EXPERIMENT 28: SMALL-SCALE REDOX TITRATION

Report Form

DATA

Calibration of a Beral Pipet

Number of drops per 3 mL:

_____ drops (trial 1)   _____ drops (trial 2)   _____ drops (trial 3)

Number of drops per 1 mL:

_____ drops (trial 1)   _____ drops (trial 2)   _____ drops (trial 3)

Average number of drops per 1 mL: _____ drops

Preparation of KMnO₄ Solution

Mass of $KMnO_4$ used: _____ g   or   Volume of 0.__ _ M $KMnO_4$ used: _____ mL

Volume of water used: _____ mL   Total Volume of diluted solution: _____ mL

Standardization

	Determination		
	1	*2*	*3*
Mass of $Na_2C_2O_4$ used:	_____ g	_____ g	_____ g
Drops of $KMnO_4$ soln. used:	_____ drops	_____ drops	_____ drops
Volume of $KMnO_4$ soln. used:	_____ mL	_____ mL	_____ mL

Unknown determinations

Unknown # _____

Determination

	1	2	3
Final buret reading:	_____ mL	_____ mL	_____ mL
Initial buret reading:	_____ mL	_____ mL	_____ mL
Volume of unknown used:	_____ mL	_____ mL	_____ mL
Drops of $KMnO_4$ soln. used:	_____ drops	_____ drops	_____ drops
Volume of $KMnO_4$ solution used:	_____ mL	_____ mL	_____ mL

Date _____   Instructor's Signature _____

ANALYSIS

Calibration of a Beral Pipet

Average number of drops per mL: _____ drops

Standardization

Molarity of the $KMnO_4$ soln.:   (determination #1) _____ *M*

(determination #2) _____ *M*

(determination #3) _____ *M*

Average _____ *M*

Std. deviation: _____

Relative std. deviation: _____ %

Record all calculations including balanced equations in your notebook and attach a copy to this Report Form. Questions on the Postlab Questions may require that you use some of these values again.

Unknown determinations

Molarity of the iron(II) soln: (determination #1)   _____ *M*

(determination #2)   _____ *M*

(determination #3)   _____ *M*

Average   _____ *M*

Std. deviation: _____

Relative std. deviation: _____ %

(Attach your calculations)

Assume the unknown iron solution was prepared by (1) dissolving 25.00 g of an impure sample of hematite (Fe_2O_3 containing ore) in hot nitric acid, (2) boiling off the excess nitric acid, dissolving the iron(III) nitrate in water, reducing the iron(III) ions to iron(II) ions, and (3) diluting the product of the 25.00-g ore sample to exactly 500 mL. From these data and your analysis of the iron(II) in solution, calculatethe percent Fe_2O_3 in the ore sample.

% Fe_2O_3 in the ore sample = _____ % (Attach your calculations)

POSTLAB QUESTIONS

1. Based upon the molarity you calculated for the unknown iron solution, how many moles of iron ion would be contained in 1.00 liter of the unknown iron ion solution? How much would that many moles of iron ion weigh?

2. What are the similarities and differences between redox titration and acid–base titration?

3. In acid–base titrations an indicator is added. Why is an indicator not used in this redox titration?

4. What are the similarities and differences between counting drops in a microscale experiment and using a buret in a macroscale experiment?

Date _____ **Student's Signature** _____

The Copper Cycle

An Open Inquiry Experiment

INTRODUCTION

You have completed a number of labs dealing with acid–base, redox, and/or displacement reactions. In those prior experiments, you have studied applications of acid–base indicators, gravimetric and volumetric stoichiometry, and chemical reactivity. This experiment is an open inquiry lab dealing with chemical reactions. "Open" means that you choose the details you wish to investigate from a number of options and you design the procedures. You will collect data that will suggest the series of reactions that you will investigate further. Your results will be used to prepare a "reaction cycle." A "reaction cycle" is a series of reactions that start with a compound or element; after several intermediate products, it eventually results in the starting material being regenerated.

OBJECTIVES

You (with the assistance of a partner) will organize a series of experiments that will yield the specific "reaction cycle"—the "copper cycle." You will test the possible reaction of metallic copper and a series of reagents, test the product of the initial reaction selected with the reagents, and repeat the selection of reactions until metallic copper is once again obtained. Compounds in the cycle are to include copper(II) hydroxide, copper(II) nitrate, and copper(II) sulfate (not necessarily in that order).

CONCEPTS

Since elemental copper is the beginning and final compound, you will begin by testing possible reactions between copper and dilute solutions of HCl, HNO_3, H_2SO_4, and NaOH. You may wish to consult the Activity Series in your textbook. The Activity Series indicates that copper will not react with acid or water to produce hydrogen gas. The Activity Series also indicates that an oxidizing compound (acid) is required to oxidize copper.

By knowing that redox occurs and that hydrogen is not a product, you can write and balance a possible equation for the reaction should any of the above dilute reactants react.

TECHNIQUES

Solution handling, using litmus paper, vacuum filtering, writing and balancing equations, and weighing are several of the techniques used in this experiment.

ACTIVITIES

You will design and test a series of reactions that will convert copper into three successive compounds before converting the last compound back to copper. This means that you will convert copper into a copper-containing compound, convert that copper compound into a different copper-containing compound, convert that compound into yet a different copper-containing compound, and then convert that compound back to copper.

> **CAUTION**
>
> **You will be working with dilute solutions but you should treat all solutions as potentially dangerous. Dispose of materials as directed by your instructor. Wear approved eye protection.**

PROCEDURES

Preliminary Tests

1-1. Design an experiment that will determine whether or not metallic copper will react with water, 3 M HCl, 6 M HNO$_3$, 6 M NaOH, or 2 M H$_2$SO$_4$.

1-2. Run tests on very small samples of copper(II) hydroxide, copper(II) nitrate, copper(II) sulfate, and copper(II) oxide. Which are soluble in water, what colors are their solutions, etc.? Test any possible reactions between these compounds and the acid or base solutions listed in Procedure 1-1.

Experimental Design

2-1. Your "copper cycle" must include copper(II) hydroxide, copper(II) nitrate, and copper(II) sulfate (not necessarily in that order). Therefore, your initial reaction should be metallic copper plus one of the acids or bases above to yield one of the salts containing oxidized copper above plus a second product that is a reduction product involving the anion from the acid or base (Cl$_2$, NO$_2$, H$_2$O, or SO$_2$).

2-2. Design a sequence of reactions that you believe will give the "copper cycle." Try to design your cycle so there will be a soluble product, then an insoluble product, and then a different soluble product.

NOTE: Add the reagents slowly with mixing. If copper(II) hydroxide is formed too rapidly, it will lose water and form the black copper(II) oxide.

2-3. The third copper salt in your "copper cycle" must be reacted with a metal capable of displacing metallic copper from its combined form. For this purpose, metallic zinc is provided. Avoid adding a large excess of zinc. Any excess zinc can be removed by carefully adding 3 *M* HCl. Some reactions may be slow. Mild warming will increase the rate of reaction.

2-4. As a final step in your proposed cycle, you will isolate the copper produced by first using the vacuum filtration apparatus and procedures described in the Common Procedures and Concepts Section at the end of this manual. While it is still in the Buchner funnel, wash the copper obtained with 3 to 5 mL of acetone and draw air through the filter for about 2 minutes to dry the copper.

2-5. Write a problem statement and proposed procedures in your lab notebook. Your proposed procedures should describe the equipment needed and the amounts of materials you plan to use. Also include the reactions (equations) expected to occur based on your preliminary tests.

The Cycle

3-1. You should now be ready to test your proposed cycle. Show your proposed cycle to your instructor and have him/her sign your Report Form.

CAUTION

If your cycle includes the reaction of copper metal and nitric acid, that reaction must be carried out in the hood. NO_2 is toxic. Use the hood for any reaction giving off fumes.

3-2. Weigh and record the mass of between 0.1 and 0.2 g of copper wire or copper turnings. Run your proposed cycle. Work in the hood when you find it necessary. Record your observations during each step of the cycle. Weigh your final product. Have your instructor sign your notebook and Report Form.

Report

4-1. After completing the data collection, you will write a lab report. Although you collect data and share ideas with a partner, you will be expected to write the final lab report independently. Your grade will depend on the thoroughness of your investigation, the presentation of your data, the careful analysis of the data, and the logic used to give reasonable results and explanations.

4-2. The lab report MUST include the following four sections:

Problem Statement: This includes a few sentences describing what specific questions you were trying to answer with your experiment.

Procedures: This section contains the materials and equipment that you actually used (these may differ from those you proposed), the type of data collected (the variables measured), and the number of trials done. Remember to discuss safety considerations. Your experimentation should take up ⅔ of the lab period.

Data/Analysis: Include the data you collected. Analysis of the data should also be included (analysis = what your data tells you). Chemical equations should be given and the connection to the data should be shown.

Conclusion: This is the generalization or explanation you have deduced from your experiment. This is also the place to make explanations for any data results that are counter to logical chemical ideas and to describe how you would change the experiment if you repeated it.

Name (Print)   Date (of Lab Meeting)   Instructor

Course/Section

29   EXPERIMENT 29: THE COPPER CYCLE

Prelab Exercises

Preliminary Tests to be run.

Proposed Procedures for the Preliminary Tests (insert more sheets if needed)

Date _____   **Student's Signature** _____

Instructor's Approval and Comments:

Date _____ **Instructor's Signature** _____

Name (Print)	Date (of Lab Meeting)	Instructor

Course/Section	Partner's Name (If Applicable)

Report Form

DATA

Record the results of your preliminary tests in your notebook. Write a problem statement and proposed procedures in your notebook. Once you have your instructor's approval, collect data in your lab notebook.

Date _____   **Instructor's Signature** _____

LAB REPORT

Attach this sheet to your lab report that includes the **PROBLEM STATEMENT** (actual), **PROCEDURES** (actual), DATA/ANALYSIS, and CONCLUSION.

POSTLAB QUESTIONS

1. The title of this experiment is The Copper Cycle. Why is it referred to as a cycle?

2. Discuss at least three of the most important factors to keep in mind when designing an experiment.

3. Draw a particle view(s) of the reaction(s) in your experiment. Add extra sheets if needed.

4. Calculate the percent yield of copper obtained from the overall process.

5. Explain how your yield for copper could be greater than 100% or less than 100%.

6. Answer any questions assigned by your instructor.

Date _____ **Student's Signature** _____

Organic Molecules

A Guided Inquiry Experiment

INTRODUCTION

Organic chemistry deals with carbon-containing compounds. A subset of carbon-containing compounds is that group of compounds containing only carbon and hydrogen (hydrocarbons). In this experiment, we will look at the names and structures of hydrocarbons and halogenated hydrocarbons.

To name a hydrocarbon, begin by drawing the Lewis structure or a skeletal representation. Starting at either end, number the carbon atoms in the longest chain of carbon atoms and choose the root name that reflects that number of carbon atoms.

# of C's Root Names	1 meth	2 eth	3 prop	4 but	5 pent	6 hex	7 hept	8 oct	9 non	10 dec

Determine the suffix by examining the bonds:

1. If only single bonds are present, the ending "ane" is added to the root name.

2. If one double bond is present, the ending "ene" is added to the root name.

3. If one triple bond is present, the ending "yne" is added to the root name.

4. If two double bonds are present, the ending "diene" is added to the root name. (An "a" is often added between the root and the suffix.)

If the hydrocarbon contains a double or triple bond, the longest chain is numbered starting at the end that will result in the lowest number for the first atom in a double or triple bond. Place the number at the beginning of the name. So $CH_3CH_2CH=CHCH_3$ is named 2-pentene. (pent- for the five carbons, -ene for the double bond, and 2 because the double bond begins with the second carbon. The lower number is obtained in this example when the carbons are numbered from right to left.)

Hydrocarbon groups that are not part of the longest chain are referred to as alkyl groups. In the experiment, we will encounter only two alkyl groups, the methyl group (-CH_3) and the ethyl group (-CH_2CH_3).

If the hydrocarbon does not contain a double or triple bond, the longest chain is numbered starting at the end that will give the smallest number to the first carbon to which an alkyl group is attached. The location and name of an alkyl group is indicated by placing, in front of the root name, the name of the alkyl group proceeded by the number of the carbon to which the alkyl group is attached. $CH_3CH_2CH_2CH(CH_3)CH_2CH_3$ is named 3-methylhexanene. (hex- for the five carbons, -ane for only the single bonds, methyl for the one carbon that is not a part of the longest chain, and 3 because the methyl group is attached to the third carbon if one numbers right to left. A larger number is obtained in this example when the carbons are incorrectly numbered from left to right.)

Cyclic (ring) groups are selected as the root if they contain more carbon atoms than any single attached group or when the cyclic group contains a double or triple bond. When the cyclic group is the root, numbering may start at any carbon in the ring and go in either direction. If the cyclic group is not the root, it may be named as an alkyl group.

Examples: is methylpentane (the location of the methyl group is understood to be 1 unless noted otherwise), the double bond containing compound CH_3 is 3-methylcyclopentane (the double bond is considered to be at carbons 1 and 2), and $CH_3CH_2CH_2CH_2-$ is 1-cyclopropyl-butane.

An organic compound that contains a halogen but not a double-bonded oxygen or nitrogen can be named as if the halogen is a hydrocarbon. In these compounds, the halogen is given the name fluoro-, chloro-, bromo-, or iodo-.

When more than one alkyl or halo group is present, the names of like groups are combined and the prefixes di-, tri-, etc., are used to indicate the number of like groups being combined. The names of the groups are arranged alphabetically in the name without regard to the prefixes di, etc.

Examples: is 1-chloro-2-methylpentane and is 3-chloro-3-methylcyclopentene.

OBJECTIVES

In this experiment, you will look for patterns as you practice writing formulas, names, Lewis structures, and skeletal representations of organic molecules. You will go from 2-dimensional representations to 3-D drawings.

CONCEPTS

This experiment uses the concepts of valence electrons, Valence Shell Electron Pair Repulsion Theory (VSEPR), hybridization, bond angles, pi and sigma bonds, and representations of molecular formulas.

ACTIVITIES

You will be assigned six compounds. Four of the compounds will come from the two groups listed in the Procedures section and two will be assigned from another source.

CAUTION

Use approved eye protection if anyone in the lab is using or moving any glassware or chemicals.

Compound List

methane	$CH_2\!=\!C(CH_3)_2$
1-chloro-2-iodocyclohexane	$CH_3(CH_2)_6CH_3$
3-methyl-2-butene	$CH_3CH(CH_3)CH_2Br$
4-methyl-2-pentyne	$CH_2\!=\!CHCH_2CH_2I$
tetrachloromethane	$CH_2\!=\!CHCH_2CH_2CH_3$
2,3-dimethylbutane	CH_3CH_2Br
1,2,3-trifluoropropane	$CH_3CH\!=\!CH_2$
1,2-dichloroethene	$ClCH_2CH_2CH_3$
1,3-dibromo-2-butene	$CH_3CH_2CH_2Br$
propyne	$FCH_2CH_2CH_2CH_3$
3-bromo-4-chloro-3-methyl-1-butyne	$CH_3CH(Cl)CH(CH_3)CH_3$
1,4-difluoropentane	$CH_2\!=\!CHCH_2CH(CH_3)_2$
methylbenzene	$CH_3CH\!=\!CHCl$
2-butene	
1-chloro-2-methyl-2-hexanene	F— hexagon
3-ethyl-1,3-hexadiene	
2,4-dichloro-2-pentene	

PROCEDURES

1-1. You will be assigned two molecules from each column in the "Compound List."

1-2. You will be assigned a fifth and a sixth compound that may not be included in the lists above.

1-3. Draw the Lewis structure showing all atoms of each of the compounds assigned. Your instructor may ask that you show the formal charges on each atom or perhaps only the net formal charge. Make a model of each of the compounds assigned. Consult your textbook for more information.

1-4. Have your model approved. Draw a 3-dimensional representation of your models. Label the bond angles.

1-5. Determine the hybridization of the carbon atoms and the presence or absence of pi bonds. (Resonance structures will not be considered in this experiment. If resonance structures are practical, one of the resonance structures will reveal the 3-D structure but perhaps not the correct bond angles, etc.)

30   EXPERIMENT 30: ORGANIC MOLECULES

Prelab Exercises

1. Describe what is meant by "the octet rule."

2. How do organic molecules differ from other molecules?

3. What is a skeletal representation of a molecular formula? Give an example of a "regular" formula and a skeletal representation.

4. Why is it important that chemists have a standard naming system for compounds?

Date _____   **Student's Signature** _____

Name (Print) _____ Date (of Lab Meeting) _____ Instructor _____

Course/Section _____ Partner's Name (If Applicable) _____

30 **EXPERIMENT 30: ORGANIC MOLECULES**

Report Form

DATA

Give the following for each ion or molecule assigned to you:

1. Compound formula	1. Compound formula
2. Compound name	2. Compound name
3. Attach Lewis structure including any net formal charges.	3. Attach Lewis structure including any net formal charges
4. 3-D model approved _____ Attach 3-D sketch	4. 3-D model approved _____ Attach 3-D sketch
5. Show the bond angles on Lewis structure.	5. Show the bond angles on Lewis structure.
6. Give the number of each type of hybridization for the carbons _____ sp^3 _____ sp^2 _____ sp	6. Give the number of each type of hybridization for the carbons _____ sp^3 _____ sp^2 _____ sp
7. Number of pi bonds present _____	7. Number of pi bonds present _____

1. Compound formula

2. Compound name

3. Attach Lewis structure including any net formal charges.

4. 3-D model approved _____
   Attach 3-D sketch

5. Show the bond angles on Lewis structure.

6. Give the number of each type of hybridization for the carbons

   _____ sp^3 _____ sp^2 _____ sp

7. Number of pi bonds present _____

---

1. Compound formula

2. Compound name

3. Attach Lewis structure including any net formal charges.

4. 3-D model approved _____
   Attach 3-D sketch

5. Show the bond angles on Lewis structure.

6. Give the number of each type of hybridization for the carbons

   _____ sp^3 _____ sp^2 _____ sp

7. Number of pi bonds present _____

---

1. Compound formula

2. Compound name

3. Attach Lewis structure including any net formal charges.

4. 3-D model approved _____
   Attach 3-D sketch

5. Show the bond angles on Lewis structure.

6. Give the number of each type of hybridization for the carbons

   _____ sp^3 _____ sp^2 _____ sp

7. Number of pi bonds present _____

---

1. Compound formula

2. Compound name

3. Attach Lewis structure including any net formal charges

4. 3-D model approved _____
   Attach 3-D sketch

5. Show the bond angles on Lewis structure.

6. Give the number of each type of hybridization for the carbons

   _____ sp^3 _____ sp^2 _____ sp

7. Number of pi bonds present _____

Date _____ **Instructor's Signature** _____

POSTLAB QUESTIONS

1. Give three or more relationships between names and structures of organic molecules that you found useful during this experiment. Give an example of each.

2. Draw two different 3-dimensional representations of $CH_3CH{=}CBrCH_3$ in which the atoms are kept in about the same locations. (Hint: Ball-and-stick, wedged-bond structures, space-filling drawings, and stereo-views are examples of molecular visualizations or 3-D representations that you should consider.)

3. Referring to the structure of $CH_3CH = CBrCH_3$ which was drawn in Postlab Question 2, which four atom combination is planar? Which six atom combination is flat (planar)?

4. Describe the part that each of the following plays in the name: 2-bromo-3-methyl-3-hexene.

   2-bromo

   3-methyl

   3- - - -ene

   hex

Date _____ **Student's Signature** _____

Reactions of Organic Compounds

A Guided Inquiry Experiment

INTRODUCTION

Organic compounds by definition contain carbon, but they can contain other atoms. The arrangement of the atoms or structure of the compound can vary. Chemists study the reactions of compounds. Certain reactions are considered desirable, for example, the ability to fight cancer. Drug companies and other researchers often examine the possible link between structure and reaction. In this experiment, you will also examine any relationship between reaction and structure.

OBJECTIVES

During this experiment, you will observe the reaction of organic compounds during five tests. Based on the test results, you will propose a relationship between the reaction and structure of a compound. You will also study the reaction of an alcohol and an organic acid.

CONCEPTS

This experiment uses the concepts of solution concentration, molarity, and Lewis structures.

TECHNIQUES

Dispensing drops, observing reactions for evidence of a chemical change, and safely smelling vapors are some of the techniques that you will use during this experiment.

ACTIVITIES

In this experiment, you will investigate the reactions of organic compounds, then group the compounds by those that react in a similar manner. The structure of compounds within each group will be compared. Finally, you will investigate the reaction of two specific organic compounds.

CAUTION

Use approved eye protection. Be aware of chemicals and hot equipment.

PROCEDURES

Results of Five Tests

1-1. Obtain small test tubes and a dropper bottle of an organic compound. (You will eventually test the 10 organic compounds shown on the following page. A glass well plate can be substituted for the test tubes.)

1-2. You will perform five tests on each of the organic compounds.

1-3. Add 10 drops of the first organic compound to the first test tube. Gently waft the vapors towards your nose. Record your results.

1-4. Add 10 drops of distilled water to the first test tube. Use your stirring rod to transfer some of the liquid to pH paper or litmus paper. Record results.

1-5. Add about 10 drops of 10% $NaHCO_3$ solution to the first test tube. Record results.

1-6. Add 10 drops of the 10% $CuSO_4$ to a second test tube. Then add 4 drops of the first organic compound. Gently tap the end of the test tube, while holding the top. Record results.

1-7. Add 10 drops of the first organic compound to a third, clean test tube. Then add 2 drops of 4% ceric ammonium nitrate (CAN) in 2 M HNO_3. Gently tap the end of the test tube, while holding the top. Record results.

1-8. Repeat Procedures 1-3 through 1-7 with each of the organic compounds. When you have finished, empty the glassware and waste products as directed by your instructor. Clean the glassware.

The Reaction of Two Organic Compounds

2-1. Obtain a small test tube. Set up a hot water bath with a small beaker containing about 1 to 2 inches of water. The water should remain just below boiling. Your test tube should be able to stand upright in the beaker.

2-2. Into the test tube, place 20 drops of saturated $C_7H_6O_3$. Then add 20 drops of CH_4O.

2-3. Add 5 to 7 drops of 9 M H_2SO_4 to the test tube in the hood. Mix by gently swirling with a stirring rod. Put the test tube in a hot water bath and let it remain there for 10 minutes.

2-4. Obtain a small beaker with about 15 mL of a saturated solution of $NaHCO_3$. Pour the contents of the test tube into the beaker and waft any fumes in the beaker toward you. Record your observations. Dispose of the contents as your instructor directs. Have your instructor sign your notebook and Report Form.

Possible Organic Compounds

$C_2H_4O_2$	C_6H_{10}	C_3H_8O

$C_4H_{11}N$	C_2H_6O	CH_2O_2

$C_2H_8N_2$	$C_7H_6O_3$ (Dissolved in 50% acetone)	CH_4O

C_5H_{10}

31   EXPERIMENT 31: REACTIONS OF ORGANIC COMPOUNDS

Prelab Exercises

1. What is the difference between a hydrocarbon and an organic compound?

2. The formula for ethanol is CH_3CH_2OH. Draw an expanded representation (ball and stick representation) of methanol. Circle the non-hydrocarbon portion of methanol.

3. Repeat question 2 for formic acid, HCO_2H.

4. What is an alkene?

5. Answer any questions your instructor assigns.

Date _____ **Student's Signature** _____

31   EXPERIMENT 31: REACTIONS OF ORGANIC COMPOUNDS

Report Form

DATA

Results of Five Tests

Write in the formula for each compound. Give your observations for each below:

	Cpd 1	Cpd 2	Cpd 3	Cpd 4	Cpd 5
Formula					
Smell					
pH					
NaHCO₃					
CuSO₄					
CAN					

	Cpd 6	Cpd 7	Cpd 8	Cpd 9	Cpd 10
Formula					
Smell					
pH					
NaHCO₃					
CuSO₄					
CAN					

The Reaction of Two Organic Compounds

Observations:

Date _____ **Instructor's Signature** _____

ANALYSIS

Results of Five Tests

What patterns do you see in the data? What substances react in a similar manner? Group the substances that react the same in one group. How many groups do you have?

Was there a similar odor for the compounds in each group? If so, describe the odor.

Look at the structure of each compound on the list after the Procedures or the one provided by your instructor. What relationship do you see between the structure and the group in which you placed the compound?

Why is the non-hydrocarbon portion of a molecule referred to as its functional group?

The Reaction of Two Organic Compounds

What evidence is there that a reaction has occurred?

Propose a reaction. There were only two products; one was water, the other is called an ester. Esters contain at least the following fragment: a carbon double bonded to one oxygen and single bonded to a second oxygen. The second oxygen is bonded to an additional carbon. Write your proposed reaction:

Draw a particle view of the reaction of the two organic compounds.

POSTLAB QUESTIONS

1. Amines are compounds that form a precipitate with CAN. What structure is present in an amine?

2. Explain the tests to determine an organic acid. Then describe the structure of an organic acid.

3. Organic compounds with —OH groups are called alcohols. Explain how you could test for the presence of an alcohol.

4. Organic compounds with a double bond between two carbons are called alkenes. Explain how you could demonstrate the difference between compounds that contain a double bond and compounds with an —OH group.

5. If a compound contains more than one functional group, what test results would you expect to observe?

6. Explain why the test tube in Procedure 2-4 was poured into a saturated solution of $NaHCO_3$. What role did the $NaHCO_3$ play?

Date _____ **Student's Signature** _____

Common Procedures
and Concepts Section

LIST OF COMMON PROCEDURES AND CONCEPTS

DATA COLLECTING

General Comments on Data Collecting

The following descriptions and guidelines for data collecting are only introductory; they may also appear in various forms in the discussion accompanying some experiments and in textbooks.

Measuring Devices

In this course you will employ a number of measuring devices such as balances, burets, graduated cylinders, pipets, thermometers, and pH meters. When you are measuring a quantity, you must select the device capable of providing the data needed. The measuring instrument selected will depend on both the accuracy and detection limit (precision) required.

If you wanted to measure a 15-mg sample of a chemical, you might think first of a top-loader or triple-beam balance because either instrument could easily give a two-decimal place weighing precision. However, the triple-beam balance is not sufficiently "sensitive." At best, it could give a mass to within ±0.02 g (20 mg). You would, therefore, require the more "sensitive" analytical balance capable of weighing to ±0.1 mg (0.0001 g). In general, the simplest measuring device having the detection limit required is selected. If you had needed to weigh "about 125 g" of a salt, the triple-beam or top-loader balance would have been a logical choice. For such a measurement, the analytical balance would have been unnecessary (and more time consuming).

DATA TREATMENT

General Comments on Data Treatment

The following guidelines and descriptions regarding data treatment, error propagation, etc., are only introductory but they should be sufficient for an introductory level college chemistry laboratory course. A more thorough treatment of statistical concepts applied to experimental data can be found in more advanced textbooks.

It is important in laboratory work to be aware of the limitations in the experimental results and the obligation of all experimenters to report their results with significant figures and to include standard deviations, percent deviations, or another form of error analysis.

Errors

The results of measurements are influenced by the limitations of the instruments used. Each measurement will have a certain error (e.g., the ±0.02 g tolerance of the triple-beam balance). If you misread the mass settings on a balance, this is not referred to as an "error." It's an "experimenter's mistake." It is assumed that your results will be limited only by instrument error. "Experimenter's mistakes" are not expected during an introductory college chemistry laboratory course.

Systematic and Random Errors

Systematic errors result when an instrument is incorrectly calibrated, when an instrument has a design flaw, or when incorrect data or methods are repeatedly used. Systematic errors affect all determinations in a set of data the same way so that the value may be precise (reproducible), but inaccurate. Random errors result from the difficulty of exactly repeating a measuring procedure. Even though skill and practice can reduce random error, it is not possible to eliminate it completely. In some cases random errors occur for reasons beyond the control of the experimenter, as in a line voltage fluctuation during the use of an electric measuring device.

All experimental results should be reported with an indication of its random error, thus expressing the precision (or reproducibility) of the results. To obtain a measure of the precision, the experiment must be repeated several (n) times. The larger n is, the better is the calculated precision. As the first step toward finding the measure of precision, the average is calculated.

Average

An average is calculated by summing the individual results and dividing this sum by the number (n) of individual values as shown for five weighings on an analytical balance.

$$\overline{X} = \frac{X_1 + X_2 + X_3 + \cdots + X_n}{n}$$

$\overline{X}$: average
X_i : individual results
n : number of measurements

1st mass	1.6752
2nd mass	1.6748
3rd mass	1.6754
4th mass	1.6747
5th mass	1.6753
Total	8.3754

$$\text{Average mass} = \frac{8.3754}{5}$$

$$= 1.67508$$

(rounded to : 1.6751)

Standard Deviation

The precision of a result is measured by the standard deviation. To calculate the standard deviation, the differences between the average and every result in a series of measurements are obtained; these differences are squared; the squares are added; and the sum of the squares is divided by the number of measurements reduced by one (n – 1). Then the square root yields the standard deviation, S, as shown for the five weighings in the calculation of an average.

$$S = \sqrt{\frac{\sum_{1}^{n} \left(X_i - \overline{X}\right)^2}{n-1}}$$

Standard Deviation for the 5 weighings:

$$S = 0.0003$$

Standard deviation calculations become tedious when n is large. Fortunately, many electronic calculators have a program that calculates averages, standard deviations, and relative standard deviations. Only the individual results must be entered.

Percent Deviation (or relative standard deviation)

For comparison of precision of several results, the percent deviation is often more convenient than the standard deviation. The percent deviation (%D) is obtained by multiplying the standard deviation by 100 and dividing by the average.

$$\% = \frac{100S}{\overline{X}}$$

Percent Deviation for the 5 weighings:

$$\%D \text{ (5 weighings)} = \frac{100 \times 0.0003}{1.6751} = 0.018\%$$

Accuracy

The accuracy (error) of a result is expressed as the difference between the true value (T) and the average from a series of measurements. True values are rarely known.

$$\text{ACCURACY (or Error)} = |T - \overline{X}|$$

The closest one can come to true values is through the use of certified standards. When a result is not very accurate, the experiment is probably troubled by systematic errors. The calculations of precision and accuracy can produce four types of results shown in the diagram. All experimenters should do their best to obtain results with good precision and good accuracy. Because true values will rarely be known to you and, in many cases, will not be available to anyone, you must base your judgment on the precision of your results. Another way of expressing precision uses a range—giving the highest and the lowest value or the difference between these values. A narrow range suggests a more reliable measurement than a broad range.

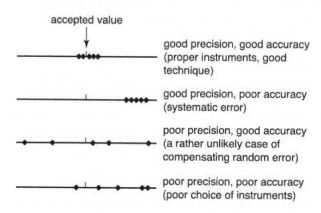

Propagation of Error

When two numbers are multiplied, divided, added, or subtracted, the precision associated with each number will effect the precision of the resulting value. For simplicity, in this course we will consider the result to have a percent deviation equal to the percent deviation of the least precise (highest percent deviation) of the two values used in the calculation.

Significant Figures

After the average and the standard deviation have been calculated, the average result must be expressed using significant figures only. In a significant-figures-only result the last figure given is in doubt. This figure is identified with the help of the standard deviation. For instance, the average of five weighings is 1.67508 g with a standard deviation of ±0.0003 g. The standard deviation tells that the fourth decimal in the result (the 0.1-mg place) is in doubt by ±3 units. Carrying the "8" in the result is not meaningful because the "0" is already uncertain and the mass could, with high probability, be anywhere between 1.6753 and 1.6747 g. If a calculation—as in the case of the five masses—yields an average with more figures than are significant, the result must be rounded to the proper number of significant figures. When the first non-significant digit is 0 through 4, omit this digit and all the following digits. When the first non-significant digit is 5 through 9, add one unit to the preceding digit.

For example : $\dfrac{1.6751}{5} = 0.33502$ (S = 0.0002) round to 0.3350

$\dfrac{1.6751}{2} = 0.83755$ (S = 0.0003) round to 0.8376

To avoid ambiguities in expressing results, use the scientific notation. Zeros to the left of the decimal point with no digits to the right of the decimal point may cause problems. When the result of a volume measurement is given as 100 mL, the two zeros are needed to fix the position of the "1" as a hundred-unit. If the two zeros were significant, for instance, in the case of a standard deviation of ±2 mL, then the result will be unambiguously expressed in scientific notation as $((1.00 \pm 0.02) \times 10^2)$ mL or at other times as 100. ± 0.02 mL.

If a measured value or a value calculated from a measured value appears without standard deviation, then the last digit of the result written in scientific notation can be assumed to be uncertain with a standard deviation of ±1. For instance, a temperature reported as 96°C can be assumed to be within 95° to 97° (96 ± 1°C).

The numbers obtained from an experiment are frequently used to calculate the final results by addition, subtraction, multiplication, and/or division. Such calculations—particularly when carried out on an electronic calculator—almost always produce more figures than are significant. How does the uncertainty (the standard deviation) in the raw data propagate to the calculated, final result? The general rule states that no mathematical operation performed on data can increase the precision. The precision is best judged in terms of the percent deviation. The final result should have the same order of magnitude as the percent deviation of the least precisely known quantity used in the calculation. Only the final result should be rounded. Earlier rounding might cause serious errors (or blunders).

An Alternate Method of Determining Significant Digits

When reporting a value that is calculated from experimental data, one often reports the final value with a number of significant digits that is based upon the following rule. The number of significant digits that is used to report an answer can be no more than the number of significant digits in the value used in the calculation that has the fewest measured digits. For introductory experimentation, both this method and the one above are usually considered and the one that yields the least precise value is generally selected.

WEIGHING TECHNIQUES

Triple-beam balances (also known as the centigram balances) and many top-loading balances can be used for rapid determination of masses to the nearest 0.01 g. These balances are considered by some users to be "rugged." However, these balances are not indestructible.

An analytical balance is a delicate instrument for the determination of masses to 0.1 of a milligram (0.0001 g). Such a balance is probably the most expensive instrument you will encounter in your laboratory. These balances will be used frequently in this course. You will need to be familiar with the proper use of all three types of balances but extra attention should be devoted to the analytical balance. Proper care and use can avoid costly repair, prevent delays due to non-functioning balances, and avoid unreliable results.

To obtain reliable results (and, perhaps, a better grade) there are a few rules that you should follow:

- **NEVER PLACE CHEMICALS DIRECTLY ON THE BALANCE PAN!!!** Place chemicals in a pre-weighed or tared beaker, etc., or on pre-weighed or tared weighing paper.
- **AVOID HANDLING PRE-WEIGHED MATERIALS.** Instead of directly touching pre-weighed materials, use a folded paper towel, tongs, etc., to avoid leaving body oil on the objects.
- **KEEP THE BALANCE AND THE AREA AROUND THE BALANCE SPOTLESSLY CLEAN!!!** If you use a balance, it is your responsibility to leave it and the area around it spotlessly clean. If you find that a balance or the surrounding area is not clean, report it to your instructor before you start to use the balance.
- **ALWAYS MAKE SURE THAT THE BALANCE IS AT THE ZERO MARK BEFORE YOU BEGIN AND AFTER YOU ARE FINISHED.** If you are unable to make it read zero, report it to your instructor.
- **DON'T DROP AN OBJECT ONTO THE PAN.** Always gently place objects onto the pan.

Weighing with a Triple-Beam Balance

The triple-beam balance consists of a pan, a supporting base, a fulcrum, a zero adjustment screw, a beam split into three sub-beams on the pointer side of the fulcrum, three mass riders, a pointer, and a fixed scale. Although you may not be able to see it, a triple-beam balance also has a magnetic dampening device.

The middle of the three sub-beams carries the 100-g rider and has notched positions in 100-g steps from 0 to 500 g. A second sub-beam carries the 10-g rider and has notched positions in 10-g steps from 0 to 100 g. The third sub-beam, equipped with a 1-g rider, has marks at 0.1-g steps from 0 to 10 g.

The mass of an object can be read to 0.1 g from the positions of the riders when the balance is at equilibrium. The position of the 1-g rider can be estimated to 0.01 g. Examine a triple-beam balance in your laboratory and identify its parts. Push the 1-g rider to an arbitrary position and practice reading the indicated mass to 0.01 g.

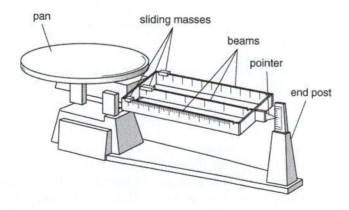

pan   sliding masses   beams   pointer   end post

Steps to follow:

1. Check the positions of the three riders. If they are not in the zero position, slide the 100-g and the 10-g riders into the zero notches and the 1-g rider exactly to the zero position. Then lightly tap the balance beam to cause the pointer to swing between the +5 and –5 positions on the fixed scale. If the balance is properly zeroed, successive pointer turning points should be at +4, –4; +3, –3; and +2, –2. When such or very similar readings are obtained, the balance is at equilibrium. If readings such as +5, –3 (or +4, –2, etc.) are obtained, ask your instructor to help you adjust the balance with the zero-adjustment screw.

2. Place the container to be weighed on the pan. The pointer will go to the top of the scale. Slide the 100-g rider into the 100-g notch. If the pointer moves down and remains down, the beaker is lighter than 100 g. Move the rider back to the zero notch. If the pointer moves up and remains up, the beaker is heavier than 100 g. In this case advance the rider one notch at a time until the pointer goes down. Then set the rider back one notch.

3. Repeat this procedure with the 10-g rider until the pointer goes down. Then set the rider back one notch. Advance the 1-g rider to the 5-g position. If the pointer moves up, set the rider to the 7.5-g position; if it moves down, to the 2.5-g position. Continue moving the rider to half-point positions until the swinging beam appears to be in equilibrium. Stop the beam; then push it lightly to have the pointer swing between the +5 and –5 scale position. If the turning points are not as described in Procedure 1, move the 1-g rider in very small increments until the beam executes the proper swings.

4. Read the rider positions and record the mass of the beaker. Read the 1-g rider position to 0.01 g.

5. Remove the container gently from the pan. Return all riders to the zero-position. Before leaving the balance, check the balance and the surrounding area. If they are not completely clean, clean up!

Weighing with a Top-Loader Balance

Top-loader balances have a pan on the top, a control area, and a digital display.

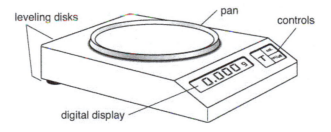

Steps to follow:

1. Turn it on; push the control to turn on the balance. After a short time, the display should read "0.00 g." It is ready for use. If it doesn't read 0.00 g or if it fluctuates greatly, push the control button again. If it still does not stabilize on zero, consult your instructor.

2. Place your beaker, watch glass, weighing paper, etc., on the pan. Wait for the reading to stabilize. Record the reading on the digital display.

3. If you wish to tare the mass of your container, push the on/off button again to tare the object on the pan. When the display stabilizes, it should read "0.00 g" again even though the object you placed on the pan is still there. Now, if a suitable container had been selected, you could carefully add material to the balance and the digital display would report only the added material.

4. Remove the objects from the pan. Allow the display to stabilize. The display should read a negative value equal to the value in step 2 above. If not, you may need to repeat the above steps.

5. Re-zero the balance. It is now ready for the next person.

Weighing with an Analytical Balance

Proper weighing on the analytical balance takes time and frequently causes "weighing jams." You will be able to use your laboratory time most efficiently when you know the balance procedures and have become an "expert weigher."

An analytical balance has the following operationally important parts: the Control Bar, the Calibration Lever, and the Balance Pan. The directions given here apply to the Mettler Model AE 200 balance. If a different model is in your laboratory, ask your instructor for any required modifications to these procedures.

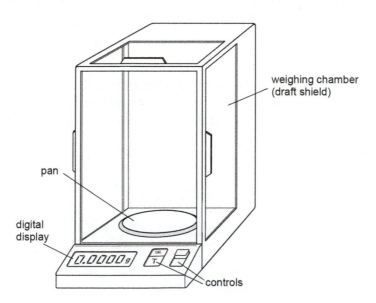

Steps to follow:

1. The balance should be located in a stable location, free of vibration. One should not move the balance once it has been given a location and calibrated.

2. The analytical balance must be kept spotlessly clean. Chemicals shall never be placed directly on the pan on any balance. Use a watch glass, a beaker, or a piece of paper to hold the chemicals. Do not attempt to weigh objects

heavier than the maximal load. Keep the doors of the balance closed unless the weighing operation requires an open door.

3. The analytical balance is switched on by briefly depressing the control bar at the front of the balance. The display will light up and show 8.8.8.8.8.8.8.8. After several seconds, the display will automatically set itself to zero.

4. If the balance has been connected to the power supply and the power has been on for at least 60 minutes, you may calibrate the balance if you have a standard 200.00 g weight. Make sure the balance doors are closed and the pan is spotlessly clean. Depress and hold the cal/menu button until the display shows "-CAL-." Then release the bar. The display will change to "- - - -" then a blinking "200.00 g." Place the standard 200.00 g weight on the pan. As soon as the display changes to "200.00," remove the weight. The balance is now ready to be used for a weighing.

5. Open one of the doors of the balance. Place your clean and dry container on the balance pan. Close the door. Wait until the display has stabilized. Read and record the mass.

6. The analytical balance has a convenient taring option. Place the container (beaker, watch glass, folded paper) on the balance pan. Close the door. Briefly press the control bar. The display changes to zero. The mass of the container is now tared. Open the door and very carefully add your chemical. Watch the display telling you the mass of chemical in the container. Add the chemical in very small increments when you approach the desired mass. When you have reached this mass, close the balance door and wait until the display has stabilized. Read and record the mass. Remove the container from the balance.

7. Before leaving the balance, make sure the weighing compartment is spotlessly clean. Close the balance doors. Do not turn off the balance unless specifically instructed to do so. If instructed to turn off the balance, lightly lift the control bar to turn it off.

BUNSEN BURNER

General Comments

The Bunsen burner is just one of several types of laboratory burners. However, it is by a large margin the type of burner most commonly found in introductory chemistry laboratories. Fisher and Tirrill burners are designed to spread the flame over a larger area, micro burners are designed to burn with a very small flame; most other burners are the Bunsen burners.

Operation

The burner must be connected to a gas source. Typically, the gas is delivered via some master valve to a petcock (gas valve) in the lab. Ask your instructor to make sure that the master valve is on. The petcock should be closed at all times except when you are using the burner. The burner will have a valve (needle valve) to control the gas flow rate. The burner also has an airflow control. The gas and air are mixed in the vertical barrel of the burner. The mixture is ignited as it emerges from the top. The proportions of air and gas are adjusted to produce the desired flame.

Steps to follow to produce the hottest possible flame:

1. Adjust the airflow so that a minimum of air is supplied.

2. Turn the gas on at both the petcock and the needle valve to give a maximum flow of gas.

NOTE: Don't screw the needle valve all the way out. If you do, turn the gas off at the petcock and replace the needle valve.

---

3. Turn the gas on and hold a lit match or striker above and to one side of the gas flow. (If you are using a striker, make it produce sparks at this time.)

4. After ignition, you will have a very yellow flame. This flame is not very hot and tends to deposit carbon on anything that it contacts. To produce a hotter flame, that does not produce a carbon deposit on glassware, slowly open the air vent. Continue until a blue inner flame is observed. The hottest part of the flame is just above the blue inner cone. If the flame blows out, turn everything off and start over, but this time don't open the air vent as much as you did the first time.

CAUTION

Never leave a flaming burner unattended. Turn everything off when you are finished.

BURETS (THEIR USE AND CARE)

A buret is a tall tube that has been calibrated to deliver, in controlled amounts, known volumes of liquids. Burets are available in a wide range of sizes and with several different types of delivery control devices such as glass or Teflon stopcocks or a simple pinchcock. The pinchcock is the most easily maintained and operated delivery control. It consists of a short piece of rubber tubing with a glass sphere stuck inside. The rubber tubing connects the buret to the glass tip. To deliver liquid from the buret, the rubber tubing is squeezed between the thumb and middle finger to form an ellipsoidal shape allowing the liquid to flow past the glass sphere.

Before using a buret, you must check whether or not the buret is clean, paying particular attention to any dirt and grease that cause water droplets to remain when water is drained from the buret. Fill the buret with tap water, dry the outside of the buret, allow the water to drain through the tip, and check for droplets clinging to the buret walls. If you observe droplets, you must clean the buret thoroughly.

Various special cleaning mixtures are sometimes employed, but most of these are dangerously corrosive. It is recommended that for cleaning a buret you use a long-handled buret brush and a detergent. Do not use abrasive cleansers that will scratch the interior of the buret and do not use soap that may leave a film on the glass. Wet the brush and sprinkle the bristles with the detergent. Insert the brush into the buret and slide it back and forth the full length of the buret 10 to 12 times. Remove the brush and rinse the buret with tap water several times. Fill the buret with tap water and drain the water through the tip. If droplets still cling to the wall, repeat the cleaning procedure. Once the buret drains cleanly, rinse it several times with distilled water. There is no need to use a lot of distilled water. Use the plastic squeeze bottle to squirt water repleatedly around the buret top to let a thin film rinse down the entire inner surface. Drain the rinses through the stopcock to rinse the stopcock and the tip.

After the buret has been cleaned, rinsed, and clamped to a ringstand, use two successive 5-mL portions of the solution to be used in the experiment to rinse the buret again. Introduce these portions with a circular motion to have them flow as a thin film down the entire inner surface. Drain the solution through the tip. Then fill the buret with the solution to above the zero mark. If bubbles cling to the sides, gently tap the outside of the buret to dislodge them. Drain enough titrant into a waste flask to expel air from the buret tip and to drop the liquid level to just below the zero mark. Read and record the liquid level.

When you are ready to measure a certain volume of solution into a container, calculate the final buret reading from the initial buret reading and the needed volume. Then open the stopcock by reaching the thumb and one of your fingers around opposite sides of the stopcock and carefully twisting it. Let the solution flow into the container. The more you twist the stopcock the faster the solution will drain into the container. When you come close to the final buret reading, slow down the delivery of solution. Add the last increments dropwise to avoid "overshooting" the required volume. When you titrate with the solution in the buret, you may add the titrant quickly at first and slow to a dropwise addition when the endpoint is near. Before reading the buret, wait a minute or two to allow liquid adhering to the walls to drain down. Then read the buret.

To read the buret, have your eye at the same level as the meniscus (curved surface) of the liquid. Record the position of the bottom of the meniscus with respect to the graduation marks on the buret. A black-and-white marked card held so that the dark line is just below (and reflected by) the bottom of the meniscus will help improve the precision of your readings. With practice you can learn to read a buret to ±0.01 mL.

When you have finished your work with the buret, drain out any remaining titrant and rinse the buret two or three times with distilled water as before. Return the buret to the storage area or, if it is to be left on the ringstand, invert it and clamp it to the ringstand. If you are asked to leave the burets filled with water, fill with distilled water and cover with an inverted test tube or a small piece of aluminum foil.

Ask your instructor for any special requirements regarding the use and care of burets in your lab.

VACUUM FILTRATION

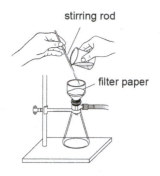

stirring rod

filter paper

A simple vacuum filtration apparatus consists of a vacuum flask (Buchner flask) with a side arm and thick glass construction, a Buchner funnel (containing a rigid support for the filter) inserted through a rubber stopper, a thick-walled vacuum hose, and a water aspirator. When water rushes through the constricted tube of the aspirator into the wider pipe, air is pulled in through the side arm of the aspirator. This airflow creates the "vacuum" for filtration.

With a water aspirator, the theoretical limit of the "vacuum" is the vapor pressure of the water at the water's temperature. In practice, particularly when a number of aspirators are running from the same water line, this limit may not be approached very closely. Filtration is accelerated by the difference between the atmospheric pressure against the surface of the liquid in the funnel and the "vacuum" pressure within the flask. A pressure difference of 730 torr is equivalent to a "push" against the liquid surface of almost one atmosphere (0.96 atm). This pressure difference is sufficient to implode a thin-walled flat-bottomed flask (such as a 250-mL Erlenmeyer). Flat-bottomed flasks (unless they have special thick glass) that are larger than 125 mL must never be evacuated.

Steps to follow:

1. Secure a clean porcelain or plastic Buchner funnel inserted through a rubber stopper to fit the vacuum flask. Set the funnel and stopper securely in the neck of the flask. Connect the flask to a water aspirator by a length of vacuum hose. To keep the flask stable, clamp it to a ringstand. Place a filter paper (of a size to fit easily in the funnel while covering all the holes in the base of the funnel) into the funnel and moisten the paper with solvent of the mixture to be filtered. Turn on the aspirator to pull the moistened paper snugly against the filter support.

NOTE: If the paper is not moistened, it might curl up when a mixture is poured into the funnel permitting part of the mixture to escape filtration.

---

2. Decant as much liquid as possible from the settled solids into the funnel. Leave as much of the solid in the beaker as possible during the initial part of the filtration. The liquid will filter more rapidly before accumulated solid begins to clog the pores of the filter paper. Do not fill the funnel more than two-thirds at any time.

3. After most of the liquid has passed through the filter, transfer the remaining liquid and solid into the funnel. Rinse the original container with small volumes of the filtrate to transfer residual solid into the funnel. Continue the aspirator vacuum at least until no further liquid drops are coming from the funnel stem. The further treatment of the solid in the

funnel will vary depending on particular experimental requirements. Stop and empty the filter flask, as described in step 4, if the flask becomes half full.

4. When the desired filtration procedures have been completed, the vacuum hose is first disconnected from the flask. Then the aspirator is turned off. If the aspirator is shut off while the hose is still connected, water may flow back into the suction flask.

TRANSFER OF SOLIDS AND LIQUIDS

Most laboratory chemicals are rather expensive. It is only good common sense, from both financial and safety standpoints, not to spill or otherwise waste chemicals. On numerous occasions you will have to transfer chemicals from stock bottles into beakers or flasks, or from one piece of equipment to another. The following suggestions will facilitate such operations.

Never take more of a chemical than you actually need. If you take too much, you cannot return the excess to the stock containers except when specifically directed to do so.

Transfer of Solids

Solids are generally stored in wide-mouth, screw-cap reagent bottles. Before taking a sample from the storage bottle, inspect the physical state of the solid in the bottle. Ideally, the solid ought to be a fine, free-flowing powder. Many solids tend to "cake." In such cases the solid particles will be stuck together. To break up such "caked" chemicals you have several options.

1. Tap the closed bottle against the palm of your hand until the "cake" has been broken up. Should tapping against the palm not produce the desired result, consult your instructor. The instructor will provide you with a completely clean and dry spatula, scoopula, or other appropriate tool with which to break up the caked chemical. This operation must be carried out with great care to avoid contamination of the chemical, breaking of the storage bottle, and injury to the hands. **Never use a glass rod to break up a solid in a storage container.**

2. To avoid contamination of a solid in a storage bottle and subsequent problems for students using such a contaminated solid, **NEVER STICK A SPATULA OR SCOOPULA INTO A STORAGE BOTTLE**. To obtain a sample of the appropriate size, take a creased piece of paper, aluminum foil, watch glass, or beaker to the storage bottle; open the bottle; bring the mouth of the bottle over your container; incline the bottle appropriately; and gently turn the bottle. The solid will flow from the bottle into your container. To transfer only slightly more than the required amount of the solid, you can perform this transfer with the container on the pan of a triple-beam or top-loader balance.

3. After you have transferred from the storage bottle to the container an amount of the solid sufficient for the experiment to be carried out, you will almost always be faced with weighing out an amount of the solid within a specified range. When the experiment calls for an amount to be weighed only to ± 0.1 g or ± 0.05 g, you can transfer the solid directly from the

storage bottle to your container on the pan of a triple-beam balance using the "turn the bottle" method described above. This transfer system will not work when the sizes of the receiving container and the storage bottle are not compatible. For instance, controlled transfer from a 5-kg storage bottle to a 100-mL beaker will be impossible and—when attempted—will cause unnecessary spillage. In such a case of mismatch, transfer first into a larger container (e.g., from the 5-kg bottle to a 500-mL beaker).

4. When you have to weigh out an amount of a solid on an analytical balance to ±0.1 mg, a creased piece of paper, a watch glass, or a beaker is placed on the balance pan. A scoopula is loaded with the solid from the container into which the solid was transferred from the storage bottle. Carefully move the scoopula over the paper, watch glass, or beaker to position the scoopula centrally about 1 cm over the container receiving the solid. Tap the hand holding the scoopula lightly with your other hand to make the solid slide off the scoopula. When you have transferred the required amount and some solid is still left on the scoopula, return the excess to your container (not the storage bottle). Then clean the scoopula with a towel.

> ### CAUTION
>
> **WHENEVER YOU SPILL ANY CHEMICAL, CLEAN UP AS SOON AS POSSIBLE. IF YOU ARE NOT SURE HOW TO CLEAN YOUR SPILL, ASK YOUR INSTRUCTOR FOR DIRECTIONS.**

Transfer of Liquids

Because liquids usually flow easily, their transfer is generally quite simple provided a few precautions are followed. Usually, you will have to pour the liquid from a larger into a smaller container. Such an operation becomes problematic when the sizes of the two containers are very different. Pouring a liquid from a 5-liter jug into a 10-cm test tube is almost impossible. In such a case, the liquid is first poured from the 5-liter jug into a 500-mL beaker, from the 500-mL beaker into a 100-mL beaker, and finally from the 100-mL beaker into the 10-cm test tube.

To pour from a large (for instance, 1 gallon) screw-cap bottle or a bottle with a flat-topped glass stopper, first remove the cap or the stopper and place it upside down on a clean towel. Then, holding your collection vessel at the mouth of the bottle, carefully tilt the bottle until liquid flows slowly and evenly into the container. Avoid rapid, jerky movements that will splatter the liquid. When you have finished, replace the cap or stopper. If any liquid has run down the outside of the container, wipe it off carefully with a damp towel.

When pouring from a glass-stoppered bottle with a "vertical disk" stopper, hold the stopper between two fingers (as shown) while pouring the liquid. Never place this type of stopper on the bench top.

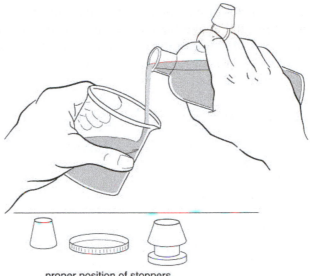

proper position of stoppers
left on bench

Special care is needed when pouring concentrated acids, concentrated bases, or solutions of corrosive or toxic reagents. Such transfers should be performed in well-ventilated hoods, with the doors of the hood drawn down as far as possible, and—if necessary—your hands protected by appropriate gloves. If you are careful, you will not spill any of these liquids. However, spills do occur. If you have a spill, you must react quickly but without panic. If you spilled some of the liquid on your hands, set the bottle and the container down in the hood and go immediately to the nearest sink. Wash the chemicals off very thoroughly with water. If you follow this procedure, you will at worst get a minor skin burn. Should you panic and drop the bottle, you and your neighbors could suffer serious harm. Consult your instructor for directions for clean up of liquids spilled in the hood.

QUALITATIVE TESTS OF COMMON GASES

Test for Water Vapor

If you suspect that a gas being produced is water and you want to test your hypothesis, first dry a strip of cobalt paper by waving it near the flame of the burner. What color is it when it is dry? You may place a very small drop of water on an edge of the paper. To test for water vapor, hold the dry cobalt paper just above the solution producing the vapors and determine if the cobalt paper turns to the color of the wet form.

Test for Hydrogen

If you suspect that a gas being produced is hydrogen and you want to test your hypothesis, first arrange to have the gas delivered to an inverted test tube filled with water held in a large pan of water. Fill the test tube half full with the gas to be tested, then slowly lift the test tube out of the water bath. Use the lighter to light a toothpick (or a splint) and hold the burning toothpick at the mouth of the still inverted test tube. The results may be startling.

> ⚠ **CAUTION**
>
> **WEAR EYE PROTECTION!**

Since hydrogen is less dense than air, an alternative method is to hold an inverted test tube above the test tube containing the reaction. If hydrogen is produced, it will rise into the top test tube. Place a lighted toothpick or splint near the mouth of the top test tube as you tilt it to release any collected hydrogen. A "bark" or small explosion is a positive test for hydrogen.

Test for Oxygen

If you suspect that a gas being produced is oxygen and you want to test your hypothesis, arrange to have the gas delivered to an inverted test tube filled with water held in a large pan of water. Fill the test tube full; then slowly lift the test tube out of the solution and invert to upright position. Oxygen gas is more dense than air. Use the lighter to light a toothpick (or splint). Blow out the flame on the toothpick. The result should be a glowing ember at the end of the toothpick. Hold the glowing toothpick at the mouth of the test tube. Use tongs to lower the glowing toothpick into the test tube. Look closely at the rate of the burning.

Since oxygen is heavier than air, an alternative method is to use the idea that any oxygen produced will stay in the bottom of the test tube containing the reaction. Place a glowing toothpick or splint into the bottom of the test tube. A glowing toothpick or splint that glows more brightly or ignites into flame is a positive test for oxygen.

Test for Carbon Dioxide

If you suspect that a gas being produced is carbon dioxide and you want to test your hypothesis, arrange to have the gas delivered to a small test tube containing a few milliliters of a clear, saturated solution of calcium hydroxide. Carbon dioxide will react with calcium hydroxide to form a white precipitate of calcium carbonate. Carbon dioxide is more dense than air.

Test for Chlorine

If you suspect that a gas being produced is chlorine and you want to test your hypothesis, first look carefully for a green colored gas being produced. To further test for chlorine, arrange to have the gas delivered to an inverted test tube half filled with water held in a large pan of water. Fill the test tube, and then slowly lift the test tube out of the solution. Very carefully smell the contents of the test tube by slowly wafting the gases at the mouth of the test tube toward your nose. Stop if you experience a smell reminiscent of bleach or a swimming pool. Chlorine gas is more dense than air.

SPECTROSCOPY

Color is one of the characteristics of a substance that may be used to determine the concentration of a solute present. When the intensity of the color can be ascertained, the concentration or quantity of the colored substance in a sample can be determined. For instance, one uses the "darkness" of the brown color of iced tea to judge whether the tea is weak or strong. If a substance does not have a color of its own, the possibility exists of reacting the colorless substance with an appropriate reagent to give a colored product.

Although the color (blue, yellow, green, red) can be easily distinguished by the human eye, the intensity of a color is much more difficult to determine. Therefore, spectrophotometers (colorimeters) were developed to electronically measure the intensity of a light beam before and after it has passed through a solution. Another device that is part of a spectrometer provides light of defined wavelengths. With such a device (a monochromator or a series of filters), the wavelengths of light that are absorbed by a solution can be determined.

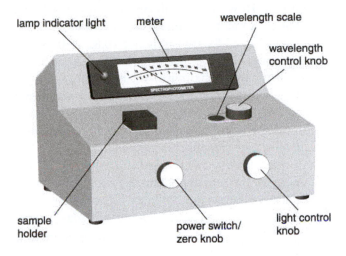

Light intensity can be measured electronically. If the incident light has an intensity of "I_o" and the transmitted light (the light passing through the sample) has an intensity of "I," then the transmittance is defined as the fraction "I/I_o." Because the values for "I" ranges from 0 (no light comes through the solution) to "I_o" (all incident light passes through), the transmittance ranges from 0 to 1. Spectrophotometers typically have scales of "% transmittance" (%T) defined as "100 I/I_o" with possible values between 0 and 100. Most spectrophotometers also have an absorbance (A) scale. Absorbance is defined as the negative logarithm of transmittance.

$$A = -\log\frac{I}{I_o} = -\log\frac{\%T}{100} = \log\frac{100}{\%T}$$

The absorption curve of a colored solution is obtained when its absorption is measured at various wavelengths and the absorbance plotted versus wavelength. Recording spectrophotometers vary the wavelength continuously and draw a continuous absorption curve. As an example, the absorption curve of an aqueous 0.10 *M* solution of copper sulfate in a 1-cm cell is shown.

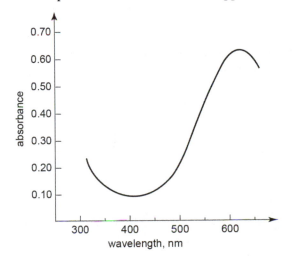

When absorbance measurements are to be used for the quantitative determination of a substance, the wavelength at which the absorption curve has a maximum is often chosen. Absorption curves may have more than one maximum. However, the absorption must not be too strong because low light intensities cannot be measured accurately. The maximum readability is between 15 to 65 % transmittance (0.82 to 0.19 absorbance). A determination in this range can be performed by picking an appropriate wavelength (not necessarily at a maximum) or by dilution of a solution that absorbs too strongly. For precise measurements at a wavelength chosen, the absorbance should not change much with a change in wavelength. This requirement is especially important when the spectrophotometer provides a band of wavelengths instead of monochromatic light.

Determining Absorption Curves in Spectroscopy

1. Turn on the spectrophotometer assigned to you and allow the instrument to warm up for several minutes. Obtain a pair of cuvets from your instructor. A cuvet is a specially made glass tube to be used with the spectrophotometer. Set the wavelength selector knob to 400 nm or the lower wavelength specified.

2. Fill one of the cuvets with distilled water. Don't touch the optical surfaces of the cuvet with your fingers. Check that no air bubbles adhere to the sides of the cuvet. If air bubbles are visible, gently shake the cuvet to dislodge the bubbles. Rinse the second cuvet with a few drops of the most concentrated solution you will be using. Then fill the cuvet with this solution.

3. With the wavelength set at 400 nm and the sample chamber empty and closed, adjust the "zero transmittance" knob until the dial reads zero transmittance. Place the cuvet of distilled water into the sample chamber. THE MARK ON THE CUVET MUST BE ALIGNED WITH THE MARK ON THE FRONT OF THE SAMPLE CHAMBER. Close the chamber cover and adjust the "100% transmittance" knob until the dial reads 100% transmittance (absorbance = 0). Repeat the zero-transmittance and 100% transmittance adjustments until the dial remains at zero (chamber closed, empty) and at 100 (cuvet of distilled water in the closed chamber).

4. Place the second cuvet (most concentrated) into the sample chamber. The mark on the cuvet must be aligned with the mark on the front of the sample chamber. Read and record the percent transmittance and absorbance.

5. Repeat the procedures in 3 and 4 until you have covered the range of wavelengths.

GRAPHING DATA

Graphing data is a good way to show the algebraic relationship between the two variables. Normally the x variable is the independent variable—the one you set. The y variable is the one that results due to the phenomena you are investigating.

- Graphs should have titles and both axes should be given a label with a unit, e.g., Mass (grams).

- Graphs should be scaled so that the data points take up at least ½ of the area of the graph, unless there is a reason otherwise.
- Each axis is scaled separately. The x axis may use gradations of 1 mL, while the y axis is in gradations of 2 g. One axis may begin at zero, the other might not. It is important to use the appropriate scaling.
- Plotting the x,y data points will give you a scatter plot from which you can look at the relationships.
- If the points look like a linear relationship exists, you can draw a best fit straight line. A best fit straight line should give a line from which data points are off the same vertical amount above the line as below the line (Δy above = Δy below).
- It is possible to disregard some data points. In this case, plot the points and explain in your write up why you felt it was valid to disregard them as experimental errors. VERY FEW points should be disregarded, and you should have a very good rationale.
- The equation for a straight line is y = mx + b, where m = slope and b = y-intercept.
- The slope is rise over run or Δy/Δx, using two points on the best fit line, NOT two data points unless they happen to be on the best fit line.
- The y-intercept is the point at which the best fit line crosses the y axis.

For example: A graph with the mass of metal pieces on the y axis and the volume of the pieces on the x, contains points that are clustered around a best fit straight line through 0,0. Why does 0,0 on a mass vs. volume graph make sense? (If there is NO volume, we have NO mass.) Since mass is on the y and volume is on the x, the equation would be mass = (slope) volume. The intercept is zero. The equation fits y = mx. The slope is mass/volume or the density. Notice that the last point was disregarded as it clearly doesn't fit the pattern of the other 5 points. It appears that the x and y data may have been recorded in the wrong column. The density here is 1.99 g mL^{-1}.

Mass (g)	Volume (mL)
2.00	3.90
5.00	12.00
3.00	5.20
6.00	12.3
8.50	15.75
8.40	4.00

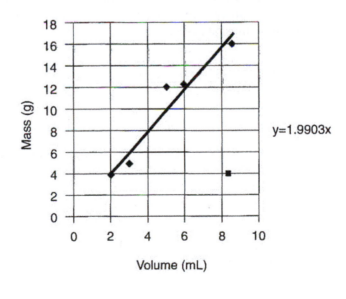

y=1.9903x

TABLE OF MOLECULAR GEOMETRIES

Regions of Electron Density	Arrangement Giving Maximum Space (electronic geometry)	Number of Regions Used in Bonding to Another Atom	Number of Lone Pairs	Molecular Geometry
2	linear	2	0	linear
		1	1	linear
3	trigonal planar	3	0	trigonal planar
		2	1	angular or bent
		1	2	linear
4	tetrahedral	4	0	tetrahedral
		3	1	trigonal pyramidal
		2	2	angular or bent
		1	3	linear
5	trigonal bipyramidal	5	0	trigonal bipyramidal
		4	1	sea saw
		3	2	T-shaped
		2	3	linear
6	octahedral	6	0	octahedral
		5	1	square pyramidal
		4	2	square planar

Note: Some books use the noun form of the geometries. For example: triangular plane, tetrahedron, triangular pyramid, triangular bipyramid, octahedron, square pyramid, square plane can be interchanged with the adjective form above.

Note: A single region of high electron density is formed by a unshared pair of electrons, a single bond, a double bond, or a triple bond.

FINDING THE EQUIVALENCE POINT

When titrating an acid with a base, one often obtains a plot of pH versus volume of base added that is similar to the line A-E in the drawing. Somewhere between points C and D, stoichiometric amounts of acid and base have been mixed, and the line changes from being a left-handed curve to being a right-handed curve. The exact point with the stoichiometric amounts is referred to as the equivalence point. We select indicators that give end points near but not always equal to the equivalence point.

To locate the equivalence point on such a plot, one can extend lines B-C and D-E to give lines B-F and G-E. A line perpendicular to the horizontal axis is drawn (line H-J). Line H-J is moved to the right or left until line C-D divides the line segment I-J into equal lengths. The point where lines H-J and C-D cross is the equivalence point. This technique for locating an equivalence point can be modified for use with titrations of bases with acids, polyprotic acids, and poly-reactive bases.

Finding the Equivalence Point

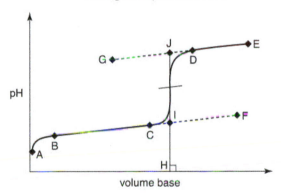